# ON THE SIGNIFICANCE OF RELIGION IN CLIMATE CHANGE

This book explores the role of religion in discussions about climate change and, particularly, the development of responses to climate change on global, state, institutional, and local levels. It considers examples of the ways that different religious traditions, including Indigenous, Muslim, Buddhist, and Christian communities, have responded to the different effects of climate change by using different methodological approaches, including political science and international relations (e.g. public opinion polls and constructivism); religious studies scholarship on climate change, including an overview of religion and ecology as a subdiscipline in religious studies; and environmental humanities approaches.

This volume interrogates the diverse ways religion acts and is acted upon by different actors, including institutions and nation states, in response to climate change. Within single traditions, different actors advocate for planetary care and concern, while their co-religionists may remain passive or deny climate change as a phenomenon.

This book hopes to complicate and unravel the complexities of how different religions approach climate change and recommends that religions are taken seriously in the development of climate change mitigation strategies at different scales.

**Lan T. Chu** is a Professor of Diplomacy and World Affairs at Occidental College.

**Amy Holmes-Tagchungdarpa** is an Associate Professor of Religious Studies and Asian Studies at Occidental College.

**Kalzang Dorjee Bhutia** is a research associate in the Hidden Stories: New Approaches to the Local and Global History of the Book project at the University of Toronto and Princeton University.

**Youssef Chouhoud** is an Associate Professor of Political Science affiliated with the Reiff Center for Human Rights and Conflict Resolution at Christopher Newport University.

## Religion Matters: On the Significance of Religion in Global Issues

Edited by Christine Schliesser, Zurich University, Switzerland, S. Ayse Kadayifci-Orellana, Georgetown University, USA and Pauline Kollontai, York St. John University, UK.

Policy makers, academics and practitioners worldwide are increasingly paying attention to the role of religion in global issues. This development is clearly noticeable in conflict resolution, development or climate change, to name just a few pressing issues of global relevance. Up to now, no book series has yet attempted to analyze the role of religion in current global issues in a coherent and systematic way that pertains to academics, policy makers and practitioners alike. The Sustainable Development Goals (SDGs) serve as a dynamic frame of reference. "Religion Matters" provides cutting edge scholarship in a concise format and accessible language, thereby addressing academics, practitioners and policy makers.

**On the Significance of Religion in Deliberative Democracy**
*Kudakwashe Chitsike, Ruby Quanston Davis, Elizabeth Gish*

**On the Significance of Religion for the SDGs: An Introduction**
*Christine Schliesser*

**On the Significance of Religion for Human Rights**
*Pauline Kollontai and Friedrich Lohmann*

**On the Significance of Religion in Climate Change**
*Lan T. Chu, Amy Holmes-Tagchungdarpa, Kalzang Dorjee Bhutia, and Youssef Chouhoud*

For more information about this series, please visit: https://www.routledge.com/Religion-Matters/book-series/RELMAT?srsltid=AfmBOoqa-UfnSCBIPKVV-C4R-a0wNy6tnlps WMJwk_uNgdCZT3iJV2bX

# ON THE SIGNIFICANCE OF RELIGION IN CLIMATE CHANGE

Lan T. Chu, Amy Holmes-Tagchungdarpa,
Kalzang Dorjee Bhutia, and Youssef Chouhoud

LONDON AND NEW YORK

Designed cover image by Taco Hammacher and Schwarzfalter.

First published 2025
by Routledge
4 Park Square, Milton Park, Abingdon, Oxon OX14 4RN

and by Routledge
605 Third Avenue, New York, NY 10158

*Routledge is an imprint of the Taylor & Francis Group, an informa business*

© 2025 Lan T. Chu, Amy Holmes-Tagchungdarpa, Kalzang Dorjee Bhutia, and Youssef Chouhoud

The right of the authors to be identified as authors of this work has been asserted in accordance with sections 77 and 78 of the Copyright, Designs and Patents Act 1988.

All rights reserved. No part of this book may be reprinted or reproduced or utilised in any form or by any electronic, mechanical, or other means, now known or hereafter invented, including photocopying and recording, or in any information storage or retrieval system, without permission in writing from the publishers.

*Trademark notice*: Product or corporate names may be trademarks or registered trademarks, and are used only for identification and explanation without intent to infringe.

*British Library Cataloguing-in-Publication Data*
A catalogue record for this book is available from the British Library

ISBN: 978-1-032-33258-1 (hbk)
ISBN: 978-1-032-33259-8 (pbk)
ISBN: 978-1-003-31886-6 (ebk)

DOI: 10.4324/b23076

Typeset in Sabon
by KnowledgeWorks Global Ltd.

# CONTENTS

*Acknowledgments*   ix

1 Summary of Implications for Academics, Policymakers, and Practitioners across and between Religious and Secular Contexts   1
   *Amy Holmes-Tagchungdarpa, Kalzang Dorjee Bhutia, Lan T. Chu, and Youssef Chouhoud*

2 Why Religion Matters for Responding to Climate Change: An Introduction   3
   *Amy Holmes-Tagchungdarpa, Lan T. Chu, Kalzang Dorjee Bhutia, and Youssef Chouhoud*

3 Relating to the Rising Waters and Warming Land: Indigenous Religions and Climate Change   12
   *Kalzang Dorjee Bhutia and Amy Holmes-Tagchungdarpa*

4 Doctrine, Praxis, and Public Opinion: Islam's Call for Environmental Stewardship and the Varied Ways It Is Answered (and Ignored)   26
   *Youssef Chouhoud*

5  Compassion and Interdependence in the Age of
   Changing Climates: Buddhist Understandings of
   Human-Environmental Relationships in the
   Anthropocene                                             44
   *Kalzang Dorjee Bhutia and*
   *Amy Holmes-Tagchungdarpa*

6  Concerned About Climate: The Catholic Church,
   Environmental Stewardship, and the
   Challenge to Brazil's Bolsonaro                          61
   *Lan T. Chu*

7  Now What? Implications for Academics,
   Policymakers, and Practitioners Across and
   Between Religious and Secular Contexts                   81
   *Kalzang Dorjee Bhutia,*
   *Amy Holmes-Tagchungdarpa,*
   *Lan T. Chu, and Youssef Chouhoud*

*Index*                                                     89

# ACKNOWLEDGMENTS

The authors would like to thank the series editors for their support throughout this process. For inspiration, the authors thank their families and each other. For material support, Lan T. Chu would like to acknowledge the John Parke Young Initiative on the Global Political Economy (Occidental College) and Kalzang Dorjee Bhutia would like to acknowledge the Robert H. N. Fellowship in Buddhist studies.

# 1

# SUMMARY OF IMPLICATIONS FOR ACADEMICS, POLICYMAKERS, AND PRACTITIONERS ACROSS AND BETWEEN RELIGIOUS AND SECULAR CONTEXTS

*Amy Holmes-Tagchungdarpa, Kalzang Dorjee Bhutia, Lan T. Chu, and Youssef Chouhoud*

Climate change, and its many effects, is leading to profound environmental and social transformations around the world. The negative impacts of these transformations are exacerbated by forms of social and environmental injustice. Religion has the significant potential to alleviate these negative impacts due to its pervasiveness. Academics, policymakers, and practitioners would benefit from interreligious and interdisciplinary dialogs and collaboration across secular and religious spheres to create sustainable and culturally nuanced pathways for the survival of the earth.

### Implication 1: Creating Conversations across Religious and Scientific Communities and Discussions

At present, scientists often resist input from religious communities and assume a binary between religious and secular action and understanding. A recognition of both scientific and religious viewpoints as sources of truth for different people can lead to discussions and collaborations based on mutual respect and increase understanding and the potential for change.

### Implication 2: Less Simplistic Understandings of Religion Are Needed

Religion operates differently across different elements of culture and society and influences how people relate to their environment. Religion itself is a complicated category and understood differently in different local and global settings. It can also be very powerful as a way to motivate action. Religions should not just be understood as representations of their institutional forms and histories, which may be tied to forms of colonialism and oppression.

## Implication 3: Personal Belief and Action Is as Important as Institutional and State Policy

Understand and uplift the power of religious actors to act as individuals instead of assuming they only act according to institutional authority. Just as religious actors do not always vote along lines set by their institutional authority, they can also make a variety of decisions without consideration of what their institutions tell them to do.

## Implication 4: Including Local Worldviews and Concerns into Decision Making in Meaningful Ways

Beyond individuals, local communities may practice their religions in different ways and have diverse understandings of their relationship with the environment. Effective policy and actions need to take into account these differences and avoid one-size-fits-all strategies.

## Implication 5: Understand How the Environment can Be Understood in Complicated Ways across Religious and Cultural Perspectives

There is no one definition of the environment, and religious actors and communities can understand the environment in different ways. These ways do not always map neatly onto the way conservationists and scientists talk about religion. Do not underestimate how this mobilization might motivate people to act in ways that conservation or environmentalist language may not capture.

## Implication 6: Listen to Indigenous Communities and Avoid Stereotypes

Indigenous communities have had longstanding relationships with the environments they live in. These relationships have been inspired by long-term observation and led to the development of innovative responses to different impacts of climate change. Incorporating Indigenous viewpoints into policy should not just include tokenism, but sustained discussions that provide Indigenous communities with agency.

## Implication 7: Climate Change Has Many Impacts – Take into Account These Different Impacts in Developing Action and Policy Beyond States

While it is important to avoid climate reductionism, climate change does have significant interconnected impacts in different social, political, and ecological contexts. Policymakers, scientists, and religious practitioners need to think about these different impacts and effects and consider holistically different responses.

# 2
# WHY RELIGION MATTERS FOR RESPONDING TO CLIMATE CHANGE

An Introduction

*Amy Holmes-Tagchungdarpa, Lan T. Chu, Kalzang Dorjee Bhutia, and Youssef Chouhoud*

**Introduction**

We are completing this manuscript as different parts of the northern hemisphere are facing their hottest summers on record. The heat is literally killing humans, nonhuman animals, and many other forms of life. However, the different discussions hosted by the United Nations Convention of Climate Change at the annual Conference of Parties on Climate Change have not yet led to the implementation of an internationally recognized set of guidelines that have been agreed upon to slow and limit the impacts of climate change. Climate change remains a controversial, widely debated topic, as misinformation remains rampant and polluting corporations still operate internationally without oversight. It is now widely known that individual decision making can make a difference, and the popularity of climate strikes and other forms of activism demonstrates that many communities are deeply concerned about climate change. However, a lack of uniform policy means that bringing people together across national boundaries to reverse climate change poses a great challenge. The internet and media have allowed for activism to spread but also for the promotion of misinformation.

Another significant element of human life that has tremendous potential to respond to climate change is religion. Both institutionally and individually, religions provide structure and guidelines to people's lives as they interact and relate to the world around them. At present, religion is all too often excluded from climate action and policy; it is dismissed as otherworldly, disengaged, or even as the source of misinformation. This book aims to provide readers with a more nuanced overview of how different religious communities and institutions are responding to the interconnected threats posed by climate change. The interdisciplinary team of authors will discuss ways that Indigenous, Christian, Buddhist, and Muslim practitioners at local and state levels conceptualize human relationships with different elements of climate change such as human and nonhuman habitat loss, environmental and meteorological change (including receding glaciers, rising sea levels, droughts, and more frequent storms), and accompanying changes to human and nonhuman health.

In this book, we consider multiple perspectives: national and international policymaking, collaboration within and across religious communities, and projects and initiatives undertaken by environmental activists. These different perspectives provide a holistic view of how communities understand their relationships to the warming planet and the impact of the changing climate on their daily lives. When it comes to conversations regarding the environment and climate change, religion usually is not the first resource to come to mind. As Hans Joachim Schellnhuber, founder and chairman of the Potsdam Institute for Climate Impact Research, noted, "Within the scientific community, there is almost a code of honor that you will never transgress the red line between pure analysis and moral issues" (Yardley and Goodstein 2015).

In this book, we go beyond scientific discussions of the harmful effects of climate change on the local, national, and international communities. We consider how religion's influence in shaping the narrative around the effects of climate change broadens the discussion to include those beyond policymakers and scientists. According to Pasqual Ferrara, religious communities play an important role in the ecological crisis because they: (1) re-frame climate change discourse into one that focuses on solidarity and equity, (2) mobilize power on behalf of Indigenous communities, (3) influence decision-making processes, and (4) develop a "hermeneutics of land" to change patterns of land use and developing human relations (Ferrara 2019, 4).

We pose the following questions that aim to connect climate change policy and religious communities: How do religious concepts such as earth stewardship in Islam and Christianity and interdependence in Buddhism and practices such as ritual and prayer act as resources for resilience as well as contributing factors in environmental degradation? How do religious lifeworlds allow people to understand change and persevere in periods of existential crises? Existing works on climate change tend to exist in disciplinary silos, where the depth and complexity of religious lifeworlds are passed over, or where detailed case studies make it difficult for general readers to access the complexity of issues at stake. In this book, we will work to explore, examine, and critically discuss religious responses to climate change with nuance and attention to different disciplinary perspectives.

## Understanding Religion

We recognize religion as a broad concept, which can manifest in different ways. To help frame our presentation of religion, we loosely adopt Frazer and Friedli's model for understanding religion's role in conflict and apply it to our approach to religion and climate change (Frazer and Friedli 2015). We specifically highlight religion as a community, a set of teachings, and as discourse. With regard to climate change, religion fosters social cohesion building to protect the environment (community), encourages state policies toward the "ought" (teachings), and pushes for a deeper understanding of how we think and act in the world (discourse). These practices can be found throughout the book, and as noted by Frazer and Friedli, "Several of these ways of thinking about religion may be relevant in any one context. They are therefore not mutually exclusive, and in many cases may be complementary" (Frazer and Friedli 2015, 6).

Furthermore, we recognize the ambivalence and diversity of single religions – that is, depending on how it is understood and used, a single religion can simultaneously motivate some toward violence and conflict and others toward peace and reconciliation or, in this context, may motivate people to care for the environment, while others remain apathetic or unmoved. While we are cognizant of religion's ability to deter or preempt efforts to protect the environment, we have chosen to highlight religion's manifestation in community, teachings, and discourse that promotes a deeper connection and concern for the environment. Similar to Scott Appleby's approach to studying nonviolent religious militants, we focus on how religious actors, in relation to climate change, "plumb their respective religious traditions for spiritual and theological insights and practices useful in preventing [environmental degradation] or limiting its spread" (Appleby 2000, 7). In utilizing this approach, we bring greater attention to religion's contributions to care for the Earth but also recognize its limitations.

## Exploring Religion and Climate Change through Multiple Perspectives

This book brings together scholarly work from the social science and humanities disciplinary perspectives of anthropology, environmental humanities, ethnic studies, history, political science, international relations, and religious studies to allow for insight into how different types of scholars approach questions around religion and climate change. Other works to date on religion and climate change have often appeared to be in disciplinary silos. For example, the collected volume *How the World's Religions Are Responding to Climate Change*, edited by Robin Globus Veldman et al. (2016), includes research predominantly carried out by geographers and development studies scholars. More recently, *Understanding Climate Change through Religious Lifeworlds*, edited by religious studies scholar David Haberman (2021), was more interdisciplinary but focused on in-depth case studies rather than overviews. Political scientists such as Pamela Chasek and David Downie have focused on the state and policymaking (Chasek and Downie 2020). Robyn Eckersley (2004) proposed a "critical political ecology", which argued for including the environment as part of the moral community, but she still focused primarily on the state's role in addressing the global environmental crisis. The recent edited volume *Climate Politics and the Power of Religion* (Berry 2022) is focused on politics and religious studies. Anthropologists Anna Tsing et al. (2017) consider how climate change will impact culture and how different residents of planet Earth across different regions of the world, including humans, nonhumans, and more-than-humans, are living together as a result of the impact of climate change.

Other works also provide insights by delving deep into specific locations or religious identities. For example, Ezra Chitando et al.'s (2022) edited volume *African Perspectives on Religion and Climate Change* provides valuable insights by considering different African nations across disciplines. Religious studies work such as Anna Gade's *Muslim Environmentalism* (2019) and Vinay Lal (2015) focuses on scholarly approaches to specific traditions, while other scholarship such as Todd LaVasseur's *Climate Change, Religion, and Our Bodily Future* (2021) considers how climate change will also impact scholarly disciplines (in the case of his book, religious studies). Jace Weaver's paper on Indigenous religious traditions provides insight into other ways people are thinking about connected traditions and their responses to climate change (Weaver 2015). Different religious organizations and scholars within these traditions are also publishing work focused on how specific religious teachings and practices can help develop environmental awareness (examples include Loy 2019; Meyaard-Schaap 2023).

Some works do consciously bridge perspectives and disciplines. This includes summary essays, such as the work by Jenkins et al. (2018) and Randolph Haluza DeLay (2014). In recent years, more scholars have demonstrated how in different communities, approaches that bridge worldviews, or accept multiple viewpoints, have been particularly effective in promoting climate change awareness (Fair 2018; Bergman 2021). This book is intended to add to these discussions, providing insight into the ways that different disciplines provide different kinds of insight into climate change and religion by incorporating perspectives from multiple traditions and locations.

**Outline of the Book**

A major challenge for examining religious conceptions of climate change and assessing responses is the tremendous diversity of religious lifeworlds. In this volume, we have chosen to examine three major religions and included the very broad category of Indigenous religions in order to represent these important perspectives that are often absent in discussions of religious traditions. In doing so, however, we also recognize the tremendous diversity within these religions, and some of them – particularly practiced by Indigenous communities – do not always label their beliefs as religious. In order to maintain diversity in this volume, each chapter also takes on a different disciplinary approach in order for the reader to gain an insight into how different types of scholarship approach the central themes of the volume.

Coming from a multidisciplinary perspective including anthropology, climate science, ethnic studies, history, Indigenous Studies, and religious studies, the first chapter examines Indigenous religions, incorporating examples from four different communities that self-identify as Indigenous. It has become a widely stated fact that Indigenous communities are disproportionately impacted by the many different impacts of climate change, including not only the warming planet but also the rise in later levels, increased flooding, and the impact on air quality and health. Across land and water, Rong and Lhopo communities in the Sikkim Himalayas, Kānaka Maoli communities in the Hawai'ian islands, and Kiowa communities of the Great Plains of North America all understand the environments in which they live, including the other nonhuman beings resident there, to be related to them in different ways. Colonial interference and interruption have attempted to disrupt these bonds of relatedness, but have not succeeded, and in recent decades, powerful forms of activism have emerged to counter the detrimental impacts of climate change and posit alternative ways of being with the earth, demonstrating the power of Indigenous environmental justice.

Just as Indigenous religions represent very different perspectives on climate change, intra-religious responses to climate change within the global Muslim community are quite varied. On the one hand, this variety should not be surprising. There are approximately two billion Muslims in the world today spread out primarily, though far from exclusively, across over 50 Muslim-majority countries. Naturally, this geographic and cultural diversity will lead to heterogeneous interpretations of a faith tradition that has evolved over 1400 years. On the other hand, the protection of the environment arguably constitutes one of these rare topics around which there is virtual consensus within Islamic jurisprudence. In both the *Quran* (Islam's holy scripture) and the *hadith* (the collected sayings, actions, and tacit approvals of the Prophet Muhammad), one finds repeated implicit and explicit calls to serve as stewards of the environment and be in balance with the natural world.

Coming from the perspective of political science, Chapter 2 considers how Islam's doctrinal unanimity around environmental issues can still manifest in very different practices and attitudes in Muslim-majority contexts. That is, as with any set of beliefs, there will always be a gap between principles and application. This chapter examines the ways in which praxis around the climate crisis manifests among Muslim political leaders and the civil society groups nested within, across, and outside Muslim-majority states. In addition, a micro-level analysis of attitudes across the Middle East and North Africa sheds light on how the Muslim population in Arabic-speaking countries regards the threat of climate change and whether religiosity drives these opinions.

Another religious tradition with global reach is Buddhism. Contemporary Buddhist organizations and conservation activists alike draw on Buddhist concepts such as interdependence, compassion, and nonviolence as inspirations for environmental movements throughout Asia and in other parts of the world. This chapter will engage with case studies from throughout Asia – from India, Taiwan, and Thailand – to explore how Buddhist communities understand and approach human-environment relations in the era of climate change. The chapter will explore multiple understandings of environmental cosmologies and forms of care (inspired by anthropologist Karin Gagné's work) beyond canonical or state narratives through an interdisciplinary lens, incorporating scholarship by anthropologists, historians, geographers, and religious studies scholars. Ultimately, just as Buddhism is varied and diverse, so too are Buddhist approaches to the changing climate and global issues that result from it.

Chapter 4 then moves back to the social sciences by focusing on religion as an institutional actor with global reach using the framework of constructivism within the discipline of international relations. Within Christianity, the understandings and approaches to climate change and the environment vary so widely, and they are beyond the scope of this book. For example, in the United States alone, scientist Thomas Ackerman identified five varied approaches among Christians: cock-eyed optimists, end-time militants, denialists, creation care proponents, and social justice advocates (Ackerman 2007, 258). As the largest Christian denomination worldwide, focusing on Catholicism – and the Catholic Church in particular – allows for a more focused, systematic approach to studying the relationship between religion and the environment.

In recent years, the role of the Catholic Church in addressing the ecological crisis was particularly evident following the 2019 presidential election of Brazil's Jair Bolsonaro. Home to the largest portion of the Amazon Rainforest, Brazil is a focal point of environmental concern. According to Carlos Nobre, an Earth scientist at the University of Sao Paulo, the Amazon also regulates global temperatures by absorbing approximately 5% of all *global* carbon dioxide emissions (Magalhaes and Pearson 2019). Considering the Earth's "single largest terrestrial carbon sink", the Amazon stores approximately absorb more carbon dioxide than it emits – i.e., 100 billion metric tons of carbon, which is more than ten times the annual global emissions from fossil fuels (Casarões and Farias 2022; Roy 2022; Greenpeace 2023, 55). Furthermore, the Amazon's billions of trees contribute to 30–50% of Brazil's rainfall and should deforestation occur in the area, a drought would ensue and South America's seasonal cycle would be affected (Chechile 2019, 36). What occurs in the Amazon, therefore, has national, regional, and global implications.

Using the framework of constructivism, the chapter examines the Catholic Church's convening of the 2007 Conference of Latin American Bishops, the 2019 Synod for the Amazon, and Pope Francis' 2015 encyclical *Laudato si'*. The confluence of these events served as a basis for those who sought to challenge Brazilian president Jair Bolsonaro's deforestation policies, which had far-reaching, deleterious effects on climate. In the language of constructivism, the Church is framed as a norm entrepreneur, who focuses mainly on prescriptive normative processes, which is distinct from other norms that establish rule-setting behaviors or constraints (Finnemore and Katherine Sikkink 1998). Using this framework, this chapter considers the structural resources and institutional design of the Catholic Church, its promotion of a human/integral ecology, and its practice of subsidiarity, which help contribute to the construction of a global norm of shared responsibility.

The final chapter considers the "Now what?" by considering the implications of themes across these chapters for academics, policymakers, and practitioners. The chapter examines the theme of shared responsibility by looking at different initiatives from within different religious communities and also across communities. These initiatives function on different scales, beginning with legal initiatives undertaken by nation states; then transnational, online contemplative initiatives; and local, grassroots initiatives. These examples all demonstrate the very different ways that religious identities and discourses have been used to inspire, and at times thwart, action against climate change. They demonstrate how the relationship between religion and climate change awareness and activism remains complicated, and the importance of continuing to engage with religious groups and discourses in response to climate change at different scales.

## Bibliography

Ackerman, Thomas. 2007. "Global Warming: Scientific Basis and Christian Responses." *Perspectives on Science and Christian Faith*. 59(4; December): 250–264.

Appleby, R. Scott. 2000. *The Ambivalence of the Sacred: Religion, Violence, and Reconciliation*. Lanham, MD: Rowman and Littlefield Publishers.

Bergman, Sigurd. 2021. *Weather, Religion, and Climate Change*. Abingdon, Oxon: Routledge.

Berry, Evan, ed. 2022. *Climate Politics and the Power of Religion*. Bloomington: Indiana University Press.

Casarões, Guilherme, and Déborah Barros Leal Farias. 2022. "Amazon and the International Order: From Promise to Peril." *Journal of International Affairs*. 75(1): 55–21.
Chasek, Pamela, and David Downie. 2020. *Global Environmental Politics*. NY: Routledge.
Chechile, Bryce. 2019. "Bolsonaro: The Fate of The Paris Climate Agreement and Global Climate Change Under a Shift in Brazilian Domestic Politics." *UC Santa Cruz Journal of International Society and Culture*. 1(1). https://doi.org/10.5070/SC61156539.
Chitando, Ezra, Ernst M. Conradie, and Susan M. Kilonzo, eds. 2022. *African Perspectives on Religion and Climate Change*. London: Routledge.
Eckersley, Robyn. 2004. *The Green State: Rethinking Democracy and Sovereignty*. Boston: MIT Press.
Fair, Hannah. 2018. "Three Stories of Noah: Navigating Religious Climate Change Narratives in the Pacific Island Region." *Geo*. 5(2): e00068.
Ferrara, Pasquale. 2019. "Sustainable International Relations. Pope Francis' Encyclical *Laudato Si'* and the Planetary Implications of "Integral Ecology." *Religions*. 10(8): 466.
Finnemore, Martha, and K. Katherine Sikkink. 1998. "International Norm Dynamics and Political Change." *International Organization*. 52: 887–917.
Frazer, Owen, and Richard Friedli. 2015. *Approaching Religion in Conflict Transformation: Concepts, Cases and Practical Implications*. Zurich: Center for Security Studies (CSS). https://css.ethz.ch/content/dam/ethz/special-interest/gess/cis/center-for-securities-studies/pdfs/Approaching-Religion-In-Conflict-Transformation2.pdf
Gade, Anna. 2019. *Muslim Environmentalisms*. New York: Columbia University Press.
Greenpeace. 2023. "Brazil and the Amazon Forest." *Greenpeace USA* (blog). https://www.greenpeace.org/usa/issues/brazil-and-the-amazon-forest/.
Haberman, David, ed. 2021. *Understanding Climate Change through Religious Lifeworlds*. Bloomington: Indiana University Press.
Haluza-DeLay, Randolph. 2014. "Religion and Climate Change: Varieties in Viewpoints and Practices." *WiREs Climate Change*. 5(2): 261–279.
Jenkins, Willis, Evan Berry, and Luke Beck Kreider. 2018. "Religion and Climate Change." *Annual Review of Environment and Resources*. 43: 85–108.
Lal, Vinay. 2015. "Climate Change: Insights from Hinduism." *Journal of the American Academy of Religion*. 83(2): 388–406.
LaVasseur, Todd. 2021. *Climate Change, Religion, and Our Bodily Future*. Lanham: Lexington.
Loy, David. 2019. *Ecodharma: Buddhist Teachings for the Ecological Crisis*. Somerville, MA: Wisdom.
Magalhaes, Luciana, and Samantha Pearson. 2019. "Brazil's Nationalism Fuels Amazon Fires — Defiant President Bolsonaro Has Strong Domestic Backing in Fight with Global Critics." *Wall Street Journal*: A.1.
Meyaard-Schaap, Kyle. 2023. *Following Jesus in a Warming World: A Christian Call to Climate Action*. Westmont: InterVarsity Press.
Roy, Diana. 2022. "Deforestation of Brazil's Amazon Has Reached a Record High. What's Being Done?" Council on Foreign Relations. August 24, 2022. https://www.cfr.org/in-brief/deforestation-brazils-amazon-has-reached-record-high-whats-being-done.
Tsing, Anna Lowenhaupt, Heather Swanson, Elaine Gan, and Nils Bubandt, eds. 2017. *Arts of Living on a Damaged Planet: Ghosts and Monsters of the Anthropocene*. Minneapolis: University of Minnesota Press.

Veldman, Robin Globus, Andrew Szasz, and Randolph Haluza-Delay, eds. 2016. *How the World's Religions are Responding to Climate Change*. New York: Routledge.
Weaver, Jace. 2015. "Misfit Messengers: Indigenous Religious Traditions and Climate Change." *Journal of the American Academy of Religion*. 83(2): 320–335.
Yardley, J, and L Goodstein. 2015. Pope Offers Radical Vision to Address Climate Change. *New York Times*: A6.

# 3
# RELATING TO THE RISING WATERS AND WARMING LAND

Indigenous Religions and Climate Change

*Kalzang Dorjee Bhutia and
Amy Holmes-Tagchungdarpa*

### Introduction

On October 4, 2023, a glacial lake overflowed in the eastern Himalayan Indian state of Sikkim and swept through the town of Chungthang, taking with it much of the town, houses, roads, bridges, and, in shocking footage, the Teesta III megadam that had been completed in 2017. As this chapter was written, new information on what contributed to the GLOF (Glacial Lake Outburst Flood) is still emerging, and the full extent of the loss of human and nonhuman animal life and damage is not known. Despite this, the GLOF has been unanimously connected to climate change in the Himalayas: the glacial lake had expanded three times in size in the last three decades (Bhushan 2023), and erratic rainfall and additional glacial melt has contributed to this. This tragedy was all the more tragic because it was forewarned by scientists monitoring the lake and also by the many diverse communities who live in Sikkim. These communities had not only warned about the glacial lake but had also protested the construction of the megadam in the first place. Anthropologist Mona Chettri has noted that the GLOF represented the coming together of two existential issues in the Himalayas: climate change and so-called development (Chettri 2023). As geographer Ritodhi Chakraborty and anthropologist Pasang Yangjee Sherpa have argued elsewhere, local communities are rarely consulted about climate change impacts in local settings (Chakraborty and Sherpa 2021). Local communities have been deeply impacted by the sense of sadness and frustration at not being heard by the state and other agencies after many years of warning.

Among these communities are groups who are Indigenous to the region, including the first people of Sikkim, the Rong (often known by their colonial

DOI: 10.4324/b23076-3

appellation, the Lepcha), and the Lhopo (or Bhutia), who are held to have migrated to Sikkim from Eastern Tibet in the thirteenth century CE. For these communities, the Teesta III and the many other dams that have been completed or are still under completion along the rivers of Sikkim pose existential threats. In both Rong and Lhopo religious cosmologies, the rivers and waterways of the Sikkim Himalayas are kin and deeply related to human beings. In Rong Indigenous cosmology, the first Rong people are held to have been made from the snow of Kanchendzonga, or in Rong language, Kongchen Konghlo, the third highest mountain in the world that presides over Sikkim, and the waterways that are connected to the glaciers on the mountain are their kin; for Rong and Lhopo Buddhists, this kinship is also recognized as the waters are recognized as protector deities that need to be cared for to sustain interdimensional wellbeing. Due to these relationships, historian Kachyo Lepcha has written that damming the rivers of Sikkim has threatened "the cradle of ... civilization" for Indigenous communities in the area, threatening biodiversity, agriculture, and other ways that people of the region relate to and live with the river (K. Lepcha 2018). Indigenous communities and their local neighbors in ethnically diverse Sikkim have been impacted by the GLOF, the erratic rainfall, irregular floods and droughts, and other manifestations of climate change in the region, which is part of the mountain range called the world's "Third Pole" since much of the world population's water comes from the Himalayan glaciers. Indigenous communities have worked with supporters in Sikkim to promote awareness of the dangers of the dams and climate change based on their longstanding relationships with, and knowledge of, Sikkim's waterways. This knowledge includes different religious lifeworlds, including Indigenous cosmologies, Buddhism, Hinduism, Christianity, and Islam; and also practices that span religions (C. Lepcha 2021). This represents one of the many places in the world where Indigenous communities have drawn on long-standing connections and relationships with the lands and waters around them to advocate against shortsighted development and for forms of well-being that are contingent on respect and reciprocity between humans and the more-than-human beings resident in the world around them.

This chapter will explore examples of how Indigenous religious lifeworlds across three settings understand, and are responding to, climate change: the Sikkim Himalayas, Hawai'i in the Pacific, and the Great Plains of North America. Here, we refer to lifeworlds as a way to acknowledge that religion can be understood as a colonial term and construct, and the terms lifeworlds and cosmologies allow for broader conceptions of culture that include, but are not limited to, religion. These communities are all now incorporated through colonialism into the contemporary nation states of India and the United States, but persist and draw on their histories and knowledge of their lands and waters to assert the potential for continued survival and wellbeing.

Sikkim experienced colonialism during the nineteenth century when British administrators from India took over the administration of the state, leading to significant cultural, political, and economic changes. In 1975, it was annexed by independent India, which has led to significant changes. Like Sikkim, Hawai'i was an independent kingdom, annexed into the United States in 1898, and the annexation continues to have a considerable impact on the islands, especially due to U.S. militarization and tourism (Silva 2004). The Kiowa people are among the different nations of the Great Plains who have been brought within the United States of America through processes of American colonialism, including genocide, "invasion, occupation, and land taking" (Rand 2008, 5).

The chapter will include discussions of secondary research from interdisciplinary perspectives, including anthropology, climate science, ethnic studies, history, Traditional Ecological Knowledge scholarship,[1] and religious studies, along with our own historical and ethnographic research in the Himalayas. Kalzang Dorjee Bhutia is a Lhopo scholar from Sikkim, and Amy Holmes-Tagchungdarpa is a Pākehā scholar from Aotearoa who has worked with Himalayan communities for over two decades. These interdisciplinary examples are not meant to be exhaustive but instead to provide a response to calls from scholars and activists to how being attentive to local forms of knowledge about the environment is crucial for truly inclusive, effective strategies to emerge to respond to climate change (Chakraborty and Sherpa 2021; Sherpa 2022a).

It has become a widely stated fact that Indigenous communities are disproportionately impacted by the many different impacts of climate change, including not only the warming planet but also the rise in later levels, increased flooding, and the impact on air quality and health. Sidelining and ignoring Indigenous knowledge about these impacts is a lost opportunity for dialog and creative solutions. As environmental scientist Jessica Hernandez has written,

> Despite climate impacts being highlighted and amplified during the pandemic of 2020, Indigenous narratives continue to be dismissed and ignored in mainstream environmentalism. The environment discourse has failed and continues to fail in uplifting and centering Indigenous peoples' voices, perspectives, and lived experiences. Ultimately, this creates further marginalization against Indigenous people and our voices, perspectives, and lived experiences are crucial and necessary to incorporate. Without them, we are continuing to separate humans from nature, and this is why our environments are at the current state they are today. It is important to note that Indigenous people have been the stewards and caretakers of our environments since time immemorial. Yet we are often left out from the environmental discourse and any decision-making pertaining to our environment.
> 
> *(Hernandez 2022, 12)*

Hernandez's argument here about the separation between humans and nature that takes place in contemporary scientific discourse has developed through a historical process. Writer Amitav Ghosh has traced this and demonstrated how the separation between humans and nature is not natural but has been constructed through the privileging of forms of science that have not been empirical and rational but are deeply entangled with colonial legacies that have sought to promote and justify capitalist extraction (Ghosh 2022).

Indigenous communities have been impacted by these legacies in complicated and detrimental ways. In the contemporary world, extraction continues, but often under the guise of development. Our argument here is not a simplistic binary – that Indigenous knowledge is inherently sustainable and scientific knowledge is bad. Instead, the communities that we highlight here all demonstrate how moving away from divisions between humans and nature provides alternative perspectives on how to live well, especially in response to climate change, and how these perspectives can engage productively with scientific discourse in mutually productive ways that are not extractive or tokenistic. We consciously avoid discussing climate change as an "emergency," since as Potawatomi philosopher Kyle Whyte has argued, colonial authorities have called events and processes "unprecedented" "emergencies," "crises," or called for "urgent" response to justify colonial intervention, violence, and harm against Indigenous people (Whyte 2021, 52–53). Instead of using these languages with their potentially destructive consequences, he encourages understanding through an "epistemology of coordination. Different from crisis, coordination refers to ways of knowing the world that emphasize the importance of moral bonds – or kinship relationships – for generating the (responsible) capacity to respond to constant change in the world" (Whyte 2021, 53). These epistemologies of coordination, which emphasize relationships between humans and nonhumans, are not an "ultimate solution," but instead point to how "education, culture, and society" can bring about important forms of change (Whyte 2021, 53).

This chapter will add to discussions about these epistemologies by centering the theme of kinship. Across land and water, Rong and Lhopo communities in the Sikkim Himalayas, Kānaka Maoli communities in the Hawaiian islands, and Kiowa communities of the Great Plains of North America,[2] all understand the environments in which they live, including the other nonhuman beings resident there, to be related to them in different ways. Colonial interference and interruption have attempted to disrupt these bonds of relatedness but have not succeeded, and in recent decades, powerful forms of activism have emerged to counter the detrimental impacts of climate change and posit alternative ways of being with the earth, demonstrating the power of Indigenous environmental justice and the significance of religious lifeways as part of these forms of justice.

## Beyond Crises, Stereotypes, and Colonial Constructs: Indigenous Religions and Environmental Knowledge

Whyte's invocation of "epistemologies" is also important, because, as this chapter will demonstrate, the term "religion" is a complicated way. All too often, "religion" has been constructed using colonial discourses, comparing different cosmologies, lifeways, and forms of knowledge against the benchmark of Christianity (Masuzawa 2005). Many Indigenous languages do not contain a word for religion. As religious studies scholar Tink Tinker has argued,

> [w]hat Indian communities have done traditionally for centuries only *becomes* religious or religion when it becomes hyper-attractive to euro-christian colonialists either for establishing colonial control or satisfying colonialist curiosities (political or academic) or to enhance their own individualist sense of religious well-being and self-empowerment (new-age seekers).
> 
> *(Tinker 2020)*

Education scholar Polly Walker has discussed the silencing of Indigenous perspectives in western academic instructions and the powerful transformative potential of including Indigenous spiritual experience and knowledge into academia (Walker 2001). Religious studies scholar Marie Alohalani Brown has posited that using the Ōlelo Hawai'i term "Ho'omana" can work as an effective alternative term to represent Hawaiian "traditional beliefs and belief-related practices" that they have fought to retain "in the face of Christian-inflected religious bias and under the hegemony of a largely Christian nation, the United States" (Brown 2022, 97). In this chapter, we take a capacious view of the idea of religion by including a wide variety of what is given different terms by different disciplinary modes of knowledge: cosmologies, traditional knowledge and practices, cultural knowledge and practices, spiritual knowledge and practices, and worldviews.

It is also important to acknowledge what is meant by Indigenous in this chapter, as this is another contested term. As a scholar of Indigenous literature Alice Te Punga Somerville has written, defining what Indigenous means is complicated, and a "simple checklist definition" can function in exclusionary and problematic ways that lead already marginalized peoples to become more so. Among the different definitions of Indigenous, including those posited by the United Nations Permanent Forum on Indigenous Issues (N.d.), "wherever they may be, agree on a few things: connection to a particular place, an experience of colonialism, and ongoing disadvantage and discrimination" (Te Punga Somerville 2023).

The communities included in this chapter have varied relationships to where they now reside: the Rong people in and around the contemporary state of Sikkim are from the region, whereas the Lhopo people are acknowledged to have migrated later. The Kānaka Maoli people of Hawai'i have long ties to the islands, and the Kiowa people record their origins as beginning in the Yellowstone River Valley in contemporary Montana and then they migrated to the Black Hills. These four communities therefore all have longstanding ties to their regions and the environments in which they lived in the past and continue to in the present.

It is therefore necessary to look past stereotypes and look at the complex relationships people have with their environments. This is especially the case when considering how religion impacts environmental perspectives. A classic article in the field of religion and ecology is historian Lynn White Jr.'s "The Historical Roots of Our Ecological Crisis" (White 1967), where he argued that Christianity "is the most anthropocentric religion the world has seen," that Christian attitudes towards the environment were shaped by a creation story that establishes man's "dominance" over nature, and that "no item in the physical creation had any purpose save to serve man's purposes" (White 1967, 1204). While even White cautioned against generalizations, this representation of Christian relationships with the environment has furthered generalizations and stereotypes about how religions more generally engage the environment and the development of a wide variety of scholarly positions, including the emergence of new religions or viewpoints entirely (Taylor 2009).

Indigenous religions have been posited as an alternative to the anthropocentrism of Abrahamic religions. However, discussing the relatedness between Indigenous people and their environments can also be problematic due to longstanding stereotypes of Indigenous people as nature worshippers. Anthropologist Shepard Krech has traced how the idea of the Ecological Indian dehumanizes Native American communities by not acknowledging the diversity and complexities of Native histories relationships with their environments (Krech 1999, 27) and the diversity of Native belief systems. The same stereotypes are also found in connection with Rong communities in the Eastern Himalayas (C. Lepcha 2021, 45) and similarly lead to ahistoricization and a flattening of diversity. This chapter will attempt to avoid these problems by using specific historical and contemporary examples connected to four Indigenous communities, and particularly the actions these communities have taken to respond to climate change and interconnected challenges that have taken place due to environmental change.

**Watery Relations**

*Mountains as Kin and Protectors in the Sikkim Himalayas*

The 2023 GLOF in Sikkim was considered particularly devastating since many different ethnic and cultural communities in Sikkim had warned that there were too many state-supported megadams in the small state (K. Lepcha

2018). Anti-dam activism began to develop in 1994, when the Sikkim Government announced plans to develop a dam on the Rathong River. At that time, diverse communities began to mobilize to protest the desecration of the river as part of West Sikkim's sacred landscape. West Sikkim was considered sacred by multiple communities and worldviews. For Rong peoples, the rivers are sacred as they flow from the snows of Kanchenjunga, which is known to Rong communities as Kinchumzongbu chyu. In Rong oral traditions, the first Rong people were created by the mother creator Itbudebu Rum from the snows of the mountain; the mountain is therefore kin, and one of the terms used by Rong for their community is *mutanchi rongcup rumkup*, which means "children of snowy peaks" (C. Lepcha 2021, 49–50). The melting of Kanchenjunga's glaciers therefore represents the disappearance of an ancestor and family member. For Lhopo, Rong, and other ethnic communities who practice Buddhism in Sikkim, Kanchenjunga is Kanchendzonga, the protector deity of the state. In both Rong Mun Bongthing traditions and Buddhist traditions that developed in Sikkim in distinct ways after Buddhism began to enter the region in the eighth century CE, ritual traditions developed to show care to Kinchumzongbu chyu/Kanchendzonga, surrounding mountains, and the foothills and valleys of Sikkim.

When the government announced the dam plan, bongthings and lamas of Pemayangtse Monastery in Sikkim invoked their ritual relationships with the mountains and waters of Sikkim. They argued that according to their cosmologies, desecration of the sacred landscape would lead to not only loss of biodiversity and impact the water but also lead to illness, poverty, and other negative consequences (Bhutia 2022, 21). In 1997, the state cancelled the project. However, since then twenty-five additional dam projects have been proposed, and many other environmental movements have emerged in opposition of these. Especially significant have been those coming from Dzongu, a Rong region in North Sikkim. As anthropologist Charisma Lepcha has vividly illustrated, in the Sikkim Himalayas – which includes areas outside of the contemporary state that are part of the broader historical Rong homeland of Mayel Liang – "water bodies are arteries of interconnectedness that connect both human and nonhuman, tangible and intangible" (C. Lepcha 2021, 44). Daily practices involving water care include many examples of folklore and ritual, both involving water and directed towards water, as well as avoiding the pollution of waterbodies since the health of waterbodies is connected with the physical and spiritual health of human residents of the land (C. Lepcha 2021, 54). Different aquatic nonhuman residents of Sikkim are also central to Mun Bongthing's ritual and medicinal practices (C. Lepcha 2021, 58). Buddhist rituals practiced by Rong, Lhopo, and other ethnic groups also hold the waters to be home to watery resident deities and spirits known as lu, which also need to be acknowledged and cared for unless they bring illness or other forms of harm (Bhutia 2022). This potential contrasts with stereotypical views of Indigenous religious lifeworlds as inherently peaceful or friendly and instead points to the complexity of relationships found in these lifeworlds.

Anthropologist Charisma Lepcha has argued that the powerful activism by groups such as Affected Citizens of the Teesta (known as ACT), which has included legal recourse, hunger strikes, and online activism undertaken in Rong communities, was often studied and represented in scholarship and media as demonstrative of Rong as "primordial environmentalists." The benefit to this has been the connection of the Rong people "with shared struggles of indigenous people across the world in a bid to safeguard their sacred waters" (C. Lepcha 2021, 45). However, these movements have also been especially significant locally, as dams distance and disturb connections between waterbodies and Rong communities and also create many other environmental problems, especially with blasting and tunneling that are part of construction and have contributed to landslides (C. Lepcha 2021, 60, 63). While the most prominent recent forms of environmental activism have been focused on dams, Lepcha's study of Rong waterviews is important as it allows us to understand how these movements are interconnected with concerns about climate change. The GLOF has been interpreted by many as an actualization of what Indigenous activists had been concerned about for decades with the installation of dams at the same time as rainfall has become more erratic and heavy, as the flooding was also caused by the heavy, erratic rainfall connected to climate change. One potential drawback to activist movements inspired by Indigenous religions is that they may not be inclusive. As anthropologist Mona Chettri has warned, sacred landscape discourse can be politicized and lead to the exclusion of non-Indigenous communities who are also adversely impacted by climate change (Chettri 2017). In the case of Sikkim, popular narratives overstate Buddhism, but other communities and religions also revere the landscape, and these different narratives of sacred landscapes can come together. The cancellation of dam projects on the Rathong River between the 1990s and 2010s demonstrated how Indigenous groups could collaborate to affect important change and raise popular awareness by drawing on concepts of relatedness with waterways.

### *Propitiating Mo'o in the Hawaiian Islands*

The importance of acknowledging human relatedness with waterways is also present beyond the mountains in ocean communities. The islands of Hawai'i are surrounded by the Pacific Ocean, and their location means that Hawaiian human and nonhuman communities are deeply experienced with the different elements of climate change: the rising of the ocean, flooding, and erratic weather patterns, among others. But freshwaters in inland places on the islands are also impacted by climate change. These are evidenced by patterns of change in how people acknowledge mo'o, a class of deities that live near or in fresh water. As Marie Alohalani Brown writes in her comprehensive study of these watery deities,

> Many mo'o have alternate forms. Predominantly female, those mo'o who masquerade as humans are often described as stunningly beautiful. Tradition holds that when you come across a body of fresh water in a secluded area and everything is eerily still, you should not linger for you have stumbled across the home of a mo'o. When the plants are yellowed and the water covered with a greenish-yellow froth, the mo'o is at home. If so, you should leave quickly lest the mo'o make itself known to you, ro your detriment. It might eat ('ai) you or take you as a lover (ai, 'to have sex')—either way, you are doomed because it will consume you completely.
>
> *(Brown 2022, 43)*

As mo'o live in water (known as wai in Hawaiian), a central part of Hawaiian culture, she argues that it is important to think of them as interconnected since mo'o have both "life-giving and death-dealing properties of wai.... They embody most is not all of its attributes" (Brown 2022, 43). Importantly, as mo'o personifies wai, both represent "continuity and regeneration" (Brown 2022, 44). Mo'o are called upon to provide water for the lands, particularly for bringing rain, and are propitiated by their human neighbors who live beside the waters through offerings.

The ability to influence and control the water is especially important in times of changing climate. Scholars of critical settler colonialism in Hawai'i Candace Fujikane have connected certain stories about the migration of the mo'o to times when

> Kānaka Maoli began to pay greater attention to the care and conservation of water and the cultivation of fish. The mo'o became known as the guardians who enforced conservation kānāwai (laws) to protect the springs, streams, and fishponds, ensuring that water was never taken for granted.
>
> *(Fujikane 2021, 2)*

Fujikane demonstrates how these forms of knowledge are part of the art of kilo, "keen intergenerational observation and forecasting key to recording changes on earth in story and song" that allow for people to see how change has met with "efforts to conserve, protect, and enhance abundance" (Fujikane 2021, 3). She outlines how kilo inspires forms of activism and care where "Kānaka Maoli are restoring the worlds where their attunement to climatic change and their capacity for kilo adaptation, regeneration, and transformation will enable them to survive what capital cannot" (Fujikane 2021, 3). Kilo is a vital Indigenous practice and an important way to promote awareness of the connections between human health and the environment that are passed down through communities outside of conventional or colonial institutions and the state.

**Landed Relations**

In the Hawaiian islands, as in Sikkim, acknowledgment of watery relations has allowed for the protection of sacred landscapes on the land as well, such as Maunakea. The water and land are interconnected. This is also the case with climate change; in the Himalayas, the receding glaciers are impacting the well-being of the soil and the growth of crops (Bhutia 2022). This connection to foodways and food sovereignty as an essential element of Indigenous well-being connects to an important example from North America, the bison.

*Relating to and Returning the Bison in Kiowa Communities*

Just as the mountains represent Himalayan communities in the popular imagination, bison are crucial to the imagined United States and are the official animal of the nation state. This is strangely ironic, given that as part of the process of constructing the nation state, bison were almost entirely eradicated as part of the broader processes of colonialism and attacks on Native sovereignty and wellbeing and specifically, as part of a project of coercion of "all Indians into compliance with the reservation system" that would ensure white access to Indian lands (Rand 2008, 69). Historically, bison were "amazingly well adapted to the arid climate of the plains …" (Rand 2008, 68) and were central to the spiritual, cultural, political, and economic lives of a number of Plains nations. For Kiowa people, bison had deep significance. They were connected with the Sun as the principal life force in the Kiowa cosmos. Bison were not only related to the Sun but also made relationships between humans and the Sun; they were considered "a mediator between Sun and the Kiowa people that could impart Sun power to fortunate power-seekers" (Kracht 2017, 74). In ritual life, bison meat was offered to the Sun as "the greatest source of power in the upper world, [that] nourishes all life forces in the lower world of the Earth" (Kracht 2017, 61). Bison body parts were used widely in healing ceremonies (Graber 2022, 130). The decline of the bison was considered an existential threat. In the 1880s, several Kiowa men prayed and carried out rituals for them to return (Graber 2022, 131–138).

By the 1890s, small communities of bison were being raised and cared for in different regions of the U.S. as different people from Native, white, and mixed backgrounds sought to save them from complete extinction with different motives. In the last decade, a spate of scientific reports and news articles have raved about bison as "climate heroes." As a story in the *Washington Post* reported,

> [t]he bison's quiet munching does more than nourish their bodies – it's one of many things they do to nurture their entire ecosystem.... Grazing bison shaving down acres of vegetarian leave more than fund behind: their aggressive chewing spurs growth of nutritious new plant shoots and their natural behaviors – the microhabitats they create by rolling in the ground, the many birds that forced symbiotic relationships with them – trickle down the food chain.
>
> *(McHugh 2022)*

The InterTribal Buffalo Council, whose work is championed in the Washington Post article, is indicative of the type of activism that has sought to return buffalo to the plains. It is now a coalition of 76 tribes from 20 states (McHugh 2022; InterTribal Buffalo Council (ITBC) 2023). As a motivator for this activity, the President of the Council, Carlson, has stated that bison are "our relatives … [t]hey're one and the same with us" and that "[t]hey belong in all of our ceremonies" (Arnold 2023). This acknowledgment and celebration of the bison seems ironic and overdue. Their ties with ecosystem health and food sovereignty initiatives (Shamon et al. 2022), together with the continued practice of Kiowa and other great plain ritual traditions, are important signs of the continued survival and efficacy of Indigenous ways of knowing.

Food sovereignty is also a central concern for Rong and Lhopo people in the Sikkim Himalayas, who are inundated with processed foods and who increasingly leave agricultural land to seek out urban employment opportunities. In Hawai'i, processed foods have been part of U.S. militarization, and there are many initiatives underway to promote Indigenous modes of food security and production (Kurashima et al. 2019). Returning to Indigenous knowledge, which emphasizes relationships with land and plants, is integral to promoting sustainable foodways into the future in these communities in regions already feeling some of the worst impacts of climate change.

## Conclusion

Indigenous religions are deeply connected with well-being across different dimensions of human and more-than-human life in different global contexts. The maintenance of different forms of well-being, including physical, spiritual, and mental health, are pressing concerns for communities who identify as Indigenous who are marginalized in a variety of ways. Anthropologist

Pasang Yangjee Sherpa has asked a key question for Indigenous communities during this era of the acceleration of climate change: "How do we live in the midst of dying?" (Sherpa 2022b). One of the potential answers explored in this chapter has been to continue to maintain relationships between humans, nonhumans, and more-than-humans, or, using Whyte's terminology, epistemologies of coordination. This maintenance continues to take place across the different communities examined in this chapter. In the eastern Himalayas, Rong and Lhopo communities continue to engage in ritual care for the waterways and the mountains; for Kānaka Maoli peoples of the Hawaiian islands, mo'o are acknowledged as co-residents of waterways; and for Kiowa communities, the buffalo connect humans with the power of the Sun.

All of these forms of knowledge and relatedness have inspired important environmental activism and movements aimed at responding to, and mitigating further damage from, climate change. These parts of religious life show a path forward for living together on a planet where the waters are rising and the mountains are turning black, and also urge response from individual, community, and state actors. The question is whether these warnings can be heeded by different parties, including at international, state, and local levels, and Indigenous sovereignties thereby honored – in Sikkim, the GLOF took place before warnings were acted upon, but the return of the buffalo to the Great Plains suggests that there is hope yet. Local communities draw on longstanding relationships with the land to respond to the many impacts of climate change, and this knowledge is crucial for creating flexible and dynamic responses within and across communities and colonially imposed state lines. Instead of tokenistic incorporation into climate responses, Indigenous communities and knowledge holders must be heard and included at all levels in the development of climate change response.

### Notes

1 For a very helpful overview of TEK in Indigenous communities in North America, see Crawford O'Brien and Talamantez (2021).
2 We do not mean to be disrespectful by using these terms for these Indigenous communities. We are aware that there are a number of terms used, including by and within these communities themselves. In addition, even the term "Indigenous" can be contested – for example, in the United States, Indigenous communities may refer to themselves as Native American, American Indian, or Indigenous. For accessibility, we have opted for widely used terms.

### Bibliography

Arnold, Billy. 2023. "Returning 'Relatives': Yellowstone sends more Buffalo to Tribes, Fewer to Slaughter." *Jackson Hole News and Guide*, July 19. https://www.jhnewsandguide.com/news/environmental/returning-relatives-yellowstone-sends-more-buffalo-to-tribes-fewer-to-slaughter/article_b834b82d-2526-52ae-b3b0-6d57ac01839e.html. (Accessed October 2, 2023).

Bhushan, Chandra. 2023. "GLOF's Sikkim Shocker." *Times of India* October 13. https://timesofindia.indiatimes.com/blogs/toi-edit-page/glofs-sikkim-shocker-much-before-the-flooding-local-green-groups-had-flagged-risks-but-mitigation-strategies-werent-a-priority-in-project-planning-goi-must-get-the-message-from/ (Accessed November 12, 2023).

Bhutia, Kalzang Dorjee. 2022. "Caring for the Land, Caring for the Dharma: The Environmental History of Buddhism at Pemayangtse Monastery, Sikkim, as a Resource for Contemporary Conservation Initiatives." In *Religion and Nature Conservation*, edited by Radhika Borde, Alison A. Ormsby, Stephen M. Awoyemi and Andrew G. Gosler. London: Routledge.

Brown, Marie Alohalani. 2022. *Ka Po'e Ko'o Akua*. Honolulu: University of Hawai'i Press.

Brown, Marie Alohalani. 2023. "Indigenous Religion of Hawai'i." *Oxford Research Encyclopedia Religion*. https://doi.org/10.1093/acrefore/9780199340378.013.1139

Chakraborty, Ritodhi, and Pasang Yangyee Sherpa. 2021. "From Climate Adaptation to Climate Justice: Critical Reflections on the IPCC and Himalayan Climate Knowledges." *Climatic Change*, 167. doi.org/10.1007/s10584-021-03158-1

Chettri, Mona. 2017. "Ethnic Environmentalism in the Eastern Himalaya." *Economic and Political Weekly*. 52(46): 34–40.

Chettri, Mona. 2023. "Who is responsible for Sikkim's glacial lake outburst flood?" *The Hindu*, October 25. https://frontline.thehindu.com/environment/article 67453490.ece (Accessed November 12, 2023).

Crawford O'Brien, Suzanne, and Inés Talamantez. 2021. "Climate and Conservation." In *Religion and Culture in Native America*. Lanham: Rowman and Littlefield.

Fujikane, Candace. 2021. *Mapping Abundance for a Planetary Future*. Durham: Duke University Press.

Ghosh, Amitav. 2022. *The Nutmeg's Curse*. Chicago: University of Chicago Press.

Graber, Jennifer. 2022. "'They Call It Ghost Dance … But It's Feather Dance': Indigenous Histories in the Study of Religion and US Empire." In *Religion and US Empire*, edited by Tisa Wenger and Sylvester A. Johnson. New York: New York University Press, 124–148.

Hernandez, Jessica. 2022. *Fresh Banana Leaves: Healing Indigenous Landscapes Through Indigenous Science*. Berkeley: North Atlantic Books.

Intertribal Buffalo Council (ITBC). 2023. *Intertribal Buffalo Council*. https://itbcbuf falonation.org (Accessed November 20, 2023).

Kracht, Benjamin R. 2017. *Kiowa Belief and Ritual*. Lincoln: University of Nebraska Press.

Krech, Shephard. 1999. *The Ecological Indian*. New York: W.W. Norton.

Kurashima, Natalie, Lucas Fortini, and Tamara Tickton. 2019. "The Potential of Indigenous Agricultural Food Production Under Climate Change in Hawai'i." *Nature Sustainability*. 2: 191–191. https://www.nature.com/articles/s41893-019-0226-1

Lepcha, Kachyo 2018. "The Teesta Hydro Power Projects: A Historical Analysis of the Protest Movement in North Sikkim (1964-2011)." PhD Dissertation, Sikkim University.

Lepcha, Charisma K. 2021. "Lepcha Water View and Climate Change in Sikkim Himalaya." In *Environmental Humanities in the New Himalayas*, edited by Dan Smyer Yü and Erik de Maaker. New York: Routledge, 43–65.

Masuzawa, Tomoko. 2005. *The Invention of World Religions*. Chicago: University of Chicago Press.

McHugh, Jess. 2022. "Once Nearly Extinct, Bison are Now Climate Heroes." *The Washington Post*, July 13. https://www.washingtonpost.com/climate-solutions/2022/07/13/bison-buffalo-oklahoma-extinct-climate-change/ (Accessed November 15, 2023).

Rand, Jacki Thompson. 2008. *Kiowa Humanity and the Invasion of the State*. Lincoln: University of Nebraska Press.

Shamon, Hila, Olivia G. Cosby, Chamois L. Andersen, Helen Augare, Johnny BearCut Stiffarm, Claire E. Brenan, Brent L. Brock, Ervin Carlson, Jessica L. Deichmann, Aaron Epps, Noelle Guernsey, Cynthia Hrtway, Dennis Jørgensen, Willow Lipp, Daniel Kinsey, Kimberly J. Komatsu, Kryan Kunel, Robert Magnan, Jeff M. Martin, Bruce D. Macwell, William J. McShea, Cristina Mormorunni, Salrah Olimb, Monica Ratling Hawk, Richard Ready, Roxann Smith, Melissa Songer, Bronch Speakthuner, Grant Stafne, Melissa Weatherwax, and Thomas S. Akre. 2022. "The Potential of Bison Restoration as an Ecological Approach to Future Tribal Food Sovereignty on the Northern Great Plains." *Fronter Ecology Evolution*. 10. https://doi.org/10.3389/fevo.2022.826282

Sherpa, Pasang Yangjee. 2022a. "Nepal's Climate-Change Cultural World." In *The Anthroposcene of Weather and Climate: Ethnographic Contributions to the Climate Change Debate*, edited by Paul Sillitoe. New York: Berghahn Books, 220–248.

Sherpa, Pasang Yangjee. 2022b. *How to Live in the Midst of Dying*. Talk on the University of California Riverside Zoom Series, Riverside: University of California.

Silva, Noenoe. 2004. *Aloha Betrayed: Native Hawaiian Resistance to American Colonialism*. Durham: Duke University Press.

Taylor, Bron. 2009. *Dark Green Religion*. Berkeley: University of California Press.

Te Punga Somerville, Alice. 2023. "Are Māori Indigenous? That's Not the Real Question." *E-Tangata*, October 8. https://e-tangata.co.nz/comment-and-analysis/are-maori-indigenous-thats-not-the-real-question/ (Accessed November 15, 2023).

Tinker, Tink. 2020. "Religious Studies – The Final Colonization of American Indians, Part 1." *Religious Theory*, June 1. https://jcrt.org/religioustheory/2020/06/01/religious-studies-the-final-colonization-of-american-indians-part-1-tink-tinker-wazhazhe-udsethe/

United Nations Permanent Forum on Indigenous Issues. N.d. "Factsheet." https://www.un.org/esa/socdev/unpfii/documents/5session_factsheet1.pdf

Walker, Polly. 2001. "Journey Around the Medicine Wheel: A Story of Indigenous Research in a Western University." *The Australian Journal of Indigenous Education*. 29(2): 18–21.

White, Lynn Jr. 1967. "The Historical Roots of Our Ecologic Crisis." *Science*. 155: 1203–1207.

Whyte, Kyle. 2021. "Against Crisis Epistemology." In *Handbook of Critical Indigenous Studies*, edited by Brendan Hokowhitu. New York: Routledge, 52–64.

# 4

# DOCTRINE, PRAXIS, AND PUBLIC OPINION

Islam's Call for Environmental Stewardship and the Varied Ways It Is Answered (and Ignored)

*Youssef Chouhoud*

**Introduction**

Muslims comprise approximately one-quarter of the world's population. A substantial portion of this global community live outside the approximately 50 Muslim-majority countries and draw (to varying degrees) on a faith tradition that has evolved over 1400 years. With spatial and temporal contours this broad, it is no surprise that there are relatively few matters garnering so little controversy that one can meaningfully state what "Islam" says on the issue. Protection of the environment arguably constitutes one of these rare topics around which there is virtual consensus. Both the *Quran* (Islam's holy scripture) and the *hadith* (the collected sayings, actions, and tacit approvals of the Prophet Muhammad) repeatedly guide believers toward actions that we would today readily code as conservationist. Indeed, there are numerous allusions to keeping care for the environment paramount even under the most dire of circumstances: during war and even during the Day of Judgement.

Of course, as with any set of beliefs, there will always be a gap between principles and application. Muslim political leaders, for example, are never solely guided by fealty to a comprehensive moral doctrine; economic, security, and social concerns often also factor into their decision-making. To be sure, these latter considerations are not entirely divorced from morality, yet they indicate a push and pull that generates a variety of outcomes in elite behavior. For Muslim non-governmental organizations engaged in environmental advocacy, it is not competing inputs but rather divergent scopes and contexts that drive variation in their impact. That is, even among those Muslim NGOs that are guided explicitly by a commitment to environmental ethics, the space for civil society in their host or target country can constrain their influence. At the micro level, too, individual preferences are similarly subject to a multitude of considerations, with religiosity being but one of the factors potentially shaping Muslim attitudes and behaviors toward climate change. This chapter outlines each of these dimensions in turn, with special emphasis on the available public opinion on climate change as a novel empirical contribution.

## The Islamic Basis for Environmental Concern

This section provides a broad (if far from comprehensive) survey of what the religious tradition of Islam has to say about a believer's duty to protect the environment. It first identifies several concepts rooted in religious terminology and relevant to environmental ethics. Thereafter, it outlines doctrinal sources that speak directly to sustainability and conservation alongside the comparatively limited set of doctrinal sources that may be interpreted to support humanity's dominion over nature. This overview then concludes by highlighting pertinent religious rulings and scholarly treatments that emphasize a Muslim's obligation to safeguard the natural world while noting the contextual factors constraining such declarations.

### Religiously Rooted Environmental Ethics

A number of terms in the Islamic lexicon carry multiple meanings, some of which are directly applicable to environmental ethics. *Khilafah*, for example, generally refers to succession or stewardship. The word "caliph" in English, denoting the leader of the Muslim nation and (as such) the successor to the Prophet Muhammad, is the anglicized derivative of this term. Within the context of the environment, it emphasizes humanity's role as caretakers of the Earth, responsible for its preservation and sustainable use. This concept underscores a believer's duty to act in a manner that is just and equitable for both current and future generations (Izzi Dien 1992). The concept of *amana*, meaning trust, similarly emphasizes the responsibility delegated by God. The earth and its resources are entrusted to humans, who are accountable for their usage and preservation. As one hadith mentions, "All of you are guardians and are responsible for your wards" (Sahih Bukhari 5200). This trust extends to the environment, making its preservation a moral and religious duty (Izzi Dien 2000).

In addition to these general articulations of ecological responsibility, there are specific terms that speak to the duty to be conscious of one's consumption. The Quran warns against excess resource depletion and waste, highlighting that such behavior is blameworthy in rather severe terms: "Indeed, the wasteful are brothers of the devils ..." (Quran 17:27). Thus, waste (*israf*) and extravagance (*tabdhir*) should be purposefully avoided when structuring one's lifestyle. Scholars have argued that these references and the behaviors they prohibit provide a basis for a sustainable and ethical approach to the environment (Izzi Dien 2000).

Along these same lines, *mizan*, meaning scale or balance, embodies the principle of moderation and proportionality. The Quran often describes the heavens and the Earth as being in a state of harmony and instructs humans not to disturb this equilibrium (Quran 55:7–8). This principle serves as a reminder to maintain balance in our consumption and use of resources, ensuring that we do not disturb the natural order (Ozdemir 2022). Scholars have also interpreted this concept as a call for sustainable environmental policies (Nasr 1996). For example, traditional Islamic agriculture practices often embody this principle by promoting crop rotation and fallow periods to keep the soil balanced and fertile.

Even concepts at the very core of Islamic theology speak to the ways in which believers should interact with the environment. *Sadaqah* refers to voluntary charity in Islam, and its implications for environmental ethics are manifold. Acts that contribute to the well-being of the Earth and its inhabitants can be considered forms of charity. For instance, the Prophet Muhammad is reported to have said, "If a Muslim plants a tree or sows seeds, and then a bird, or a person or an animal eats from it, it is regarded as a charitable gift for him" (Sahih Bukhari 2320). With references like this in mind, modern scholars extend the principle of *sadaqah* to include environmental conservation efforts and protection of endangered species (Khalid 2019).

Similarly, the principle of *'adl* (justice) in Islam also has significant implications for how Muslims conceptualize the consequences of humans' imposition on nature. Climate change is often a matter of environmental justice, as the people who contribute least to the problem are often the most affected by its impacts (Schlosberg and Collins 2014). The Islamic commitment to justice can therefore motivate action to address these disparities and to work toward climate solutions that are fair and equitable (Khalid 2002).

### *Direct References in Primary Sources*

In addition to the general alignment of several theological concepts with ecological consciousness, there are also doctrinal sources that specifically promote conservation, wildlife protection, and sustainability. Given that the earliest Muslims lived in the Arabian desert, it is not surprising that the Prophet Muhammad would advise them against being wasteful: "Do not waste water, even if you perform your ablution on the banks of an abundantly-flowing river" (Sunan Ibn Majah 425). Such reminders accord with numerous Quranic prescriptions, as well: "… Do not waste, for Allah does not like the wasteful" (Quran 6:141). One of the most oft-cited hadiths when speaking of the environment echoes one's responsibility to the natural world: "The Earth is green and beautiful, and God has appointed you his stewards over it" (Sahih Muslim 2742).

This stewardship is not only in reference to flora but to fauna as well. Care for animals is emphasized in the Quran, for example, through verses that caution against anthropocentrism: "There is not an animal that lives on the earth, nor a being that flies on its wings, but they form communities like you" (Quran 6:38). Several hadiths elaborate on the minimal level of respect and empathy that one is to show to other living creatures. For example, the Prophet Muhammad admonished that "[w]hoever kills a sparrow or anything bigger than that without a just cause, Allah will hold him accountable on the Day of Judgment'" (Sunan an-Nasa'i 4446). Elsewhere, the reward for caring for animals is emphasized: "Whoever is merciful even to a sparrow, Allah will be merciful to him on the Day of Judgment'" (Al-Adab Al-Mufrad 381) and "There is a reward for serving any living being" (Sahih Bukhari 2466).

Overall, Muslims are tasked with living a sustainable lifestyle. This approach was modeled by the Prophet Muhammad, who sewed his own clothes and mended his own shoes to avoid overconsumption. Tending to the environment was similarly encouraged: "Whoever plants a tree and diligently looks after it until it matures and bears fruit is rewarded" (Al-Musnad). This general advice applied even in the most extreme circumstances: "Even if the end of time is upon you and you have a seedling in your hand, plant it" (Musnad Ahmad ibn Hanbal, V, 415).

Of course, while the vast majority (one may even say the near totality) of doctrinal sources in Islam clearly underscore a responsibility to protect the environment, there are a select few that may be interpreted in a manner that goes against this general consensus. In several verses, the Quran refers to the authority humanity has over nature: "And He has subjected to you whatever is in the heavens and whatever is on the earth, all from Him" (Quran 45:13); "And it is He who subjected the sea for you to eat from it tender meat and to extract from it ornaments which you wear" (Quran 16:14). Though rarer, similar sentiments may be found in hadith collections as well: "Whoever revives barren land, it belongs to him" (Sunan Abu Dawood, Book 24, 3070). Broadly speaking, these references mirror the Biblical basis for humanity's hierarchy in the natural world: "Be fruitful and multiply, and fill the earth and subdue it; and have dominion over the fish of the sea and over the birds of the air and over every living thing that moves upon the earth" (Genesis 1:26).

While the verses and hadith cited above appear to grant humanity broad leeway to make use of the environment as they see fit, such an interpretation would only be viable if these references were analyzed in isolation. Read holistically, the authority that Islam's doctrinal sources grant to humanity over the natural world is more of a constrained and contingent authority rather than boundless dominion—a vice-regency, as it were, rather than absolute sovereignty. As Seyyed Hossein Nasr succinctly puts it:

> Even when in the Qur'an it is stated that Allah has subjected (*sakhkhara*) nature to man ... this does not mean the ordinary conquest of nature.... Rather, it means the dominion over things which man is allowed to exercise only on the condition that it is according to Allah's laws and precisely because he is Allah's vicegerent on earth ... this power is seen in the traditional Islamic perspective to be limited in normal circumstances by the responsibilities which he bears not only towards Allah and other men and women, but also towards the whole of creation. The Divine Law (*al-Shari 'ah*) is explicit in extending the religious duties of man to the natural order and the environment.
>
> *(Nasr 1996, 124–125)*

### *Secondary Religious Scholarship*

Armed with a cache of scriptural and prophetic references in support of environmental protection, Muslim scholars have produced public talks, official rulings, and collective declarations urging climate change action. The timing of these productions, however, can be less a function of religious or even individual factors but, depending on the location, could be reflective of state priorities and constraints.

In English-speaking countries, religious leaders often have the most freedom to craft their messages without any constraint. It is therefore not uncommon to find headline speakers in Canadian and American Muslim conventions orating passionately about the climate crisis (see, e.g., From Symbolic to Systematic in Response to Climate Change [2020]; Quran Inspires Environmental Stewardship [2015]). By that same token, one can find numerous examples of sermons that accompany the *Jumua* (Friday) prayer not only in mosques but also posted on YouTube (see, e.g., Caring for the Environment [2020]; Islam and the Environment [2021]). Indeed, there are several campaigns that encourage particular days on which mosques collectively commit to speaking about the Islamic basis for combating climate change as part of annual awareness campaigns (see, e.g., Great Big Green Week 2023 (2023)).

The same freedoms, however, are not enjoyed by religious leaders in other parts of the world. This is particularly the case in the Middle East. As Nathan Brown notes, "When believers pray in the Arab world, the state often asserts its presence .... Ministries of religious affairs generally oversee the staffing, maintenance, and operation of mosques" (Brown 2017). This oversight frequently extends to guidance on what questions from the public the central religious authorities can respond to, as well as the dissemination of "officially approved" Friday sermons.

Egypt since the 2014 military takeover is a seminal case in point when it comes to state intervention in religious affairs, which is all the more relevant given the global prominence of its religious institutions (in particular, Al-Azhar University, established in 970 CE). Again, it is not that Islamic rulings on the environment are "controversial"—rather, in the context of autocratic rule, controlling and shaping a supporting narrative (particularly through the medium of religious discourse) is imperative for regime survival. For instance, there was no mention of climate change or global warming in any of the documents published by *Dar-el-ifta'* (the official body that issues religious rulings and provides guidance on religious matters) through 2020. Indeed, the first *fatwa* (religious ruling) from this governmental body on the topic of climate change (specifically, prohibiting environmentally harmful practices) was in 2022 (Aman 2022). Not incidentally, this ruling was part of a charter disseminated one month before Egypt was to host the United Nations Climate Change Conference (COP27). This clearly choreographed action underscores the role that political leaders in Muslim-majority countries play in both framing mass attitudes and affecting mass behavior on climate change.

**Political Elites**

A unique set of dynamics underscores the importance of political leaders in Muslim-majority populations' response to climate change. First, the dearth of democracy across these countries puts added weight on the proclamations and actions of each state's leader. Of the 57 member states in the Organisation of Islamic Cooperation (OIC), only one is rated as "free" (Guyana, where Islam is notably not the majority religion) and 19 others as "partly free," according to Freedom House's 2022 "Freedom in the World" report. Since most Muslim-majority states are ruled by autocrats, their dictates are often the main (if not the sole) driver of public policy in these countries—including climate policy.

Second, while the populations of each of these countries do not meaningfully contribute to annual global $CO_2$ emissions (only Indonesia [729 metric tons—mt, 6th], Iran [691 mt, 7th], Saudi Arabia [663 mt, 9th], and Turkey [436 mt, 14th] rank in the top 20 emitters worldwide), a number of OIC member states nonetheless play a critical role in the release of carbon into the atmosphere. Specifically, many of the world's top producers of fossil fuels are Muslim-majority countries, mainly (though not exclusively) in the Middle East and North Africa. This standing holds in regards to both oil and natural gas—11 of the top 20 producers in the world for each category are members of the OIC (Energy Institute 2023). Moreover, while the burning of coal emits more carbon than either oil or gas, the latter two fossil fuels combined account for the majority of $CO_2$ released into the atmosphere each year (Friedlingstein et al. 2023).

Last, countries across the MENA will be among those that feel the effects of climate change most acutely. Indeed, the region is warming at a rate much faster than the global average—a number of cities have already seen temperature spikes north of 50 degrees Celsius (122 degrees Fahrenheit) during the summer months. If this area of the world stays on its current course, it will likely exceed 4 degrees Celsius above the pre-industrial average temperature by 2050, which will render many parts of the Arabian Peninsula virtually uninhabitable (Vohra 2023).

Left unchecked, this climate trajectory will have a devastating effect on one of the most fundamental rights in Islam: the hajj pilgrimage. Every year, around two million Muslims visit the city of Mecca in Saudi Arabia to perform the hajj rituals, thus posing a tremendous logistical challenge under the best of circumstances. However, when the hajj is held in summer (the Islamic calendar follows lunar cycles, and so the start of hajj is eleven days earlier each year than the year before), the already elevated health risks associated with a gathering that large are all the more dire. Indeed, already in 2024, hundreds of pilgrims died as a direct result of heat-related stroke and dehydration.

Of course, climate change will present MENA countries with a number of pressures beyond soaring temperatures. Desertification, biodiversity loss, water scarcity, sea level rise, and climate migration may increasingly afflict a number of states in the region (Keynoush 2023). These challenges are going to be all the more taxing due to the comparatively high population growth rates in the MENA region (outpaced only by sub-Saharan Africa). Given these dynamics, the rhetoric, planning, and actions of political leaders in Muslim-majority countries offer important insights into the context within which Islam can influence the attitudes and behaviors toward climate change.

### *Proclamations and Plans*

Leaders from a number of Muslim-majority countries have issued statements and introduced ambitious initiatives in recent years aimed at combating climate change. Such proclamations are often issued in the lead up to or during the proceedings for the annual United Nations Climate Change Conferences (also known as the Conference of Parties or simply COP). As world leaders were gearing up to attend COP26 in Glasgow, for example, there was "a flurry of regional announcements from Turkey, Saudi Arabia, and the United Arab Emirates (UAE) signaling their intention to become zero-carbon economies by the middle of the century" with "[o]ther countries across the MENA region ... [raising] their 2030 targets including Morocco, Tunisia, Lebanon, Jordan, Oman, Qatar, and Sudan" (Elgendy 2021). This regional trend was only bolstered by the hosts of 2022's COP27 and 2023's COP28 being Egypt and the United Arab Emirates, respectively. In his speech during the opening session of the 2022 conference, Egypt's President Sisi challenged those in attendance to go beyond mere rhetoric:

> Today, what our world needs to overcome the current climate crisis and to reach what we have agreed on as goals in the Paris Agreement, surpasses slogans and words. Today, our peoples expect from us rapid, effective and equitable implementation. Our peoples expect us to take real and concrete steps towards reducing emissions, enhancing adaptation with the consequences of climate change, and providing the necessary financing for developing countries that suffer the most from the current climate crisis. Therefore, we have been keen to call this Conference: "Implementation Summit," which is the goal that all our efforts and endeavors must center around.
>
> *(State Information Service 2022)*

Yet, for all these grand visions, Muslim-majority countries in the MENA and elsewhere have, by and large, fallen short of their promises to date. This reality was clearly evident during 2023's COP28 meeting when ministers from the United Arab Emirates, the host nation, were put on the defensive following the publication of a report that was highly critical of the country's plans to expand oil and gas production (Abnett 2023). Indeed, measured against the commitments set forth in the Paris Agreement, which are meant to hold global warming to well below 2 degrees Celsius above the pre-industrial average, the UAE, Saudi Arabia, Iran, Indonesia, and Turkey have each taken "critically insufficient" action, while Egypt fairs only slightly better, with their actions rated as "highly insufficient" (Climate Action Tracker 2023a).

### Actions Taken

These critiques are not applicable across all the OIC member states, however. Morocco, for example, has already taken several meaningful steps toward their climate goals:

> Rabat has turned the crisis into an opportunity and has become one of the pioneers in climate policy, not only in the region but globally. Ambitious plans were years ago put into action to secure Morocco's future. The government implemented several national strategies to enhance water management, bolster the use of non-conventional water resources and modernize irrigation systems. It has also invested heavily in constructing dams to move water from northern to arid southern regions.
>
> *(Falk 2021)*

Like Morocco, Nigeria also earns a comparatively high rating of "almost sufficient" from the Climate Action Tracker project, which notes that the country has already adopted new methane guidelines and is unveiling a new carbon tax policy in the near future (Climate Action Tracker 2023b). With Nigeria being Africa's top oil producer and Morocco being an Arabic-speaking monarchy, these countries demonstrate that structural conditions are not insurmountable when it comes to sustainable climate policy.

In each case, political leaders showed they had the vision to act in their state's interest and the wisdom to not tether their efforts exclusively to the goal of climate change mitigation. The importance of this latter point cannot be overestimated as a major obstacle to climate action is the presence of up-front costs for long-term (and potentially marginal) gain. By introducing the prospect of immediate benefits, leaders can drastically curb the inertia that stifles meaningful climate change policy. In the case of Morocco, the King was able to advance his country's international standing through climate diplomacy (Falk 2021). With regards to Nigeria, political leaders there were able to institute a climate security approach to conflict prevention (Climate for Peace 2023). Both cases also highlight a general pattern across Muslim-majority states wherein these countries' political leaders rarely couch their commitment to sustainability—whether in terms of rhetoric or tangible action—in religious terms. The same, however, cannot be said of civil society organizations in these contexts.

**Civil Society**

While there are certainly organizations that tackle climate change from a secular, exclusively scientific perspective across the Middle East and many other Muslim-majority states, the integration of Islamic principles with climate action is much more organic among civil society groups than political elites. NGOs and Islamic organizations seek to influence the discourse and policy around climate change at the local, national, and international levels. Notably, however, those entities based in Western settings are generally at the vanguard of these efforts.

Islamically rooted environmental initiatives at the local and national level are constrained (and, as a result, comparatively rare) in Muslim-majority countries but are increasingly active in North America and Western Europe. In Canada, for example, EnviroMuslims seeks to "educate and empower the Canadian Muslim community to embed sustainability in our everyday lives." (Enviromuslims 2023). This mission includes partnerships with other faith-based agendas to launch projects such as "Greening Canadian Mosques," which aims to "embed sustainability in [mosques'] operations and provide them with the tools they need to identify, track, and deliver resource efficiency opportunities" (Greening Canadian Mosques 2023).

Similar intra-communal and inter-faith efforts characterize the work of organizations in the US and the United Kingdom. Green Muslims, for example, is a non-profit based in Washington, DC, whose mission is to "serve as a source in the Muslim community for spiritually-inspired environmental education, action, and reflection" and whose vision, in line with the tenets outlined earlier in this chapter, is "to see Muslims living in the environmental spirit of Islam, striving towards connection with nature and environmental stewardship" (Green Muslims 2023). The organization thus seeks not only to encourage Muslims to become more engaged with the natural world but also to facilitate their involvement in climate activism locally and nationally.

Across the Atlantic Ocean, the Islamic Foundation for Ecology and Environmental Sciences (IFEES) serves as an active and invaluable resource for Muslims in Britain and across the globe (IFEES 2023). Within the UK, this non-profit drives a number of initiatives, from training rooted in Islamic teachings to retrofitting mosques with environmentally sustainable infrastructure. This organization was also at the forefront of an effort that ultimately led to a first-of-its-kind joint declaration. At a gathering that brought organizations and leaders from across the Muslim world together, the attendees voiced their support for a united front in the struggle to protect the environment—and, in the process, highlighted how international advocacy can offer a way around domestic precarity.

While still facing challenges at home, Muslim organizations and religious leaders are freer to voice their environmental concerns without fear of retaliation from autocratic political leaders when serving as part of a transnational effort. For example, the Islamic Declaration on Global Climate Change, issued by Muslim scholars and leaders in 2015, called on members of the *ummah* (the Muslim nation) to take a leading role in addressing climate change. The declaration also called for the phasing out of greenhouse gas emissions and a transition to 100% renewable energy (Islamic Declaration on Global Climate Change 2015).

Islamic organizations have also been involved in climate change advocacy at the United Nations. For example, Islamic Relief Worldwide has been active in the UN Framework Convention on Climate Change process, advocating for climate justice and the rights of vulnerable communities (Islamic Relief Worldwide 2020). However, the role of Islamic organizations in climate change policy is not without challenges. These organizations often face resource constraints and may struggle to make their voices heard in policy debates dominated by more established international entities. They also face the challenge of translating Islamic environmental ethics into concrete policy proposals. Nonetheless, Islamic organizations continue to play a central role in the global discourse on climate change. Whether this discourse trickles down to the masses, however, is another issue.

## Public Opinion

There are multiple dimensions that constitute public opinion around the issue of climate change. One can, for example, study to what degree individuals "believe" in climate change (either in terms of whether it is occurring or whether humans are responsible for the observed warming across the globe). An additional, though not necessarily mutually exclusive, strand of research examines respondents' views on the severity of climate change. Ideally, researchers would want to gauge both of these attitudes within the same study so that they could differentiate between respondents who do not think climate change is a serious problem because they do not believe it is occurring from those who believe in man-made climate change but are more sanguine about its severity. A third common focus of public opinion research on climate change deals with the behavior that individuals take on their own, as well as the policies they would like enacted by their elected officials. Largely due to data limitations, the following analysis will almost exclusively focus on the extent to which respondents believe that climate change is a serious problem.

There are a number of constraints when using cross-national data; two in particular are relevant to the following analysis. First, there is the issue of coverage. Even the most ambitious and well-established project of its kind, the World Values Survey (which surveys at least 1200 respondents across approximately 60 countries), may not have the particular country you are interested in researching or the particular *type* of country. In this case, there are no cross-national survey projects dedicated to recording the opinions of citizens in Muslim-majority states. Luckily, there has been a growth in regional public opinion projects over the past two decades, with one of them—the Arab Barometer—focusing on a region that provides the largest concentration of Muslim-majority states.

The second major challenge is finding a survey wave that asks questions that align with the constructs you wish to test. With regard to climate change, there are two survey waves in the Arab Barometer that ask relevant questions. The seventh and most recent wave (fielded from October 2021 to July 2022) features a number of questions that reference climate change, though none that ask about attitudes regarding the phenomenon directly. For example, Q546 asks, "Do you think our national government should do more, less, or about the same amount as it is doing right now to deal with climate change?" (Arab Barometer 2022), a question that would need a respondent to not only have an opinion on climate change but also an opinion on what should be done about it and whether their country measures up to those standards. Similarly, the other questions in this wave pit climate change against other environmental and economic issues to gauge policy priorities. In the fifth wave (fielded from 2018 to 2019), however, respondents were asked, "How serious a problem do you think the following issues are: Is

[INSERT ITEM] a very serious problem, a somewhat serious problem, not a very serious problem, not at all a serious problem?" with "climate change" being one of the items (Arab Barometer 2019). This question, given that it isolates attitudes toward climate change, forms the backbone of the following original analysis.

*Methodology*

With a measure of attitudes toward climate change in hand, the next task is settling on how to gauge religiosity. Scholars can (and do) reasonably disagree on how to best capture this multifaceted construct—that is, which attributes should be tapped from the three "Bs" (belief, behavior, and belonging), and what is the best way to gauge them both individually and collectively. For researchers working off a secondary dataset, they are largely left at the mercy of the questions that the project's PIs thought most worthy (and the wording for those questions that they thought most valid). Occasionally, one might not find any useful questions in the dataset that measure a key predictor of interest (a not-infrequent occurrence for those who study religion and politics). Luckily, with regards to the Arab Barometer (given the region of focus), one need not worry about a lack of variables gauging religious belief and practice.

As no single variable can capture the complexity of a construct as rich as religiosity, the analysis uses a scale comprised of four relevant, interrelated questions:

Q609: In general, would you describe yourself as religious, somewhat religious, or not religious?
Q609a: How often do you pray? [Never, At least once a month, Once a week, Several times a week, Once a day, Five times a day]
Q610_8: Do you always, most of the time, sometimes, rarely, or never [Pray fajr on time]?
Q610_6a: Do you always, most of the time, sometimes, rarely, or never [Listen to or read the Quran daily]?

Rather than a simple additive index, which assumes that each item contributes equally to the latent measure (in this case, religiosity), I computed factor scores for each respondent and rescaled this new variable from 0 to 1 for ease of interpretation. The full regression model for each country also includes a set of control variables to better isolate the effect of religiosity on attitudes toward climate change: economic precarity, personal security, and belief that personal economic position will be better in the future than it is currently, college education, whether the respondent has children, whether the respondent is married, whether the respondent is employed, age, and sex.

## Results

Beginning with a descriptive breakdown, Figure 4.1 shows how each country's respondents view the severity of climate change. Majorities across nearly all the countries surveyed (with the exception of Kuwait) believe climate change to be either a "somewhat serious" or "very serious" problem. Although we cannot dismiss alternative explanations, it is hard not to think that Kuwait's lucrative fossil fuel revenues, which contribute to over a quarter of the small country's GDP, are a strong influence on its citizens' attitudes toward climate change. Yet, the picture is not altogether rosy aside from this oil-rich nation.

A troublingly large proportion of respondents across all countries in the sample remain skeptical of the environmental and economic hazards that a changing climate brings. This skepticism is all the more concerning given that, again, this region is already experiencing myriad negative effects from global warming and will be among the hardest hit as these dangers increase in frequency, magnitude, and scope. Furthermore, the respondents' incredulity is not an artifact of this particular survey. A poll fielded two years after the Arab Barometer's fifth wave found that, among the 31 countries surveyed from across the globe, Egypt and Saudi Arabia were the only two where at least one-fifth of respondents were "not at all worried" about climate change (Leiserowitz et al. 2021, 9). Along these same lines, an earlier Pew study found that publics in the Middle East were less concerned about climate change than those of any other region (Drake 2013).

Does religious observance augment Muslims' concern over climate change? Figure 4.2 plots the predicted probability of reporting that climate

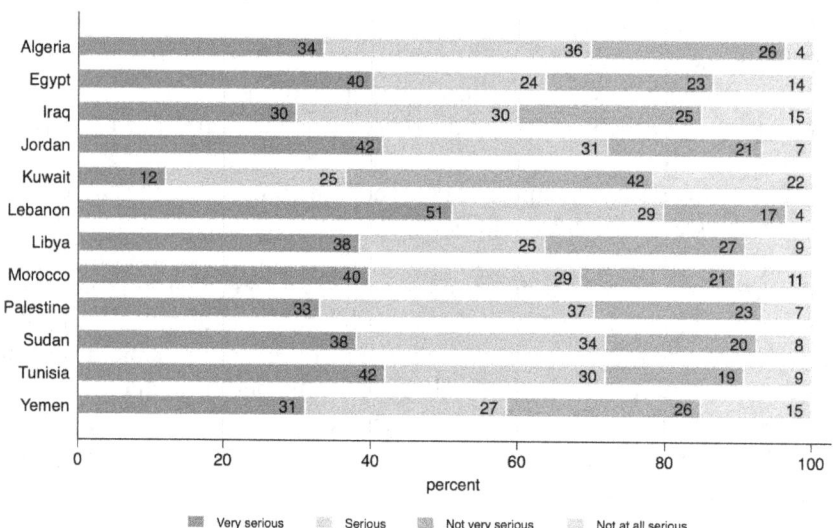

FIGURE 4.1  Descriptive analysis of Arab Barometer (Wave V) data on the seriousness of climate change

Doctrine, Praxis, and Public Opinion  39

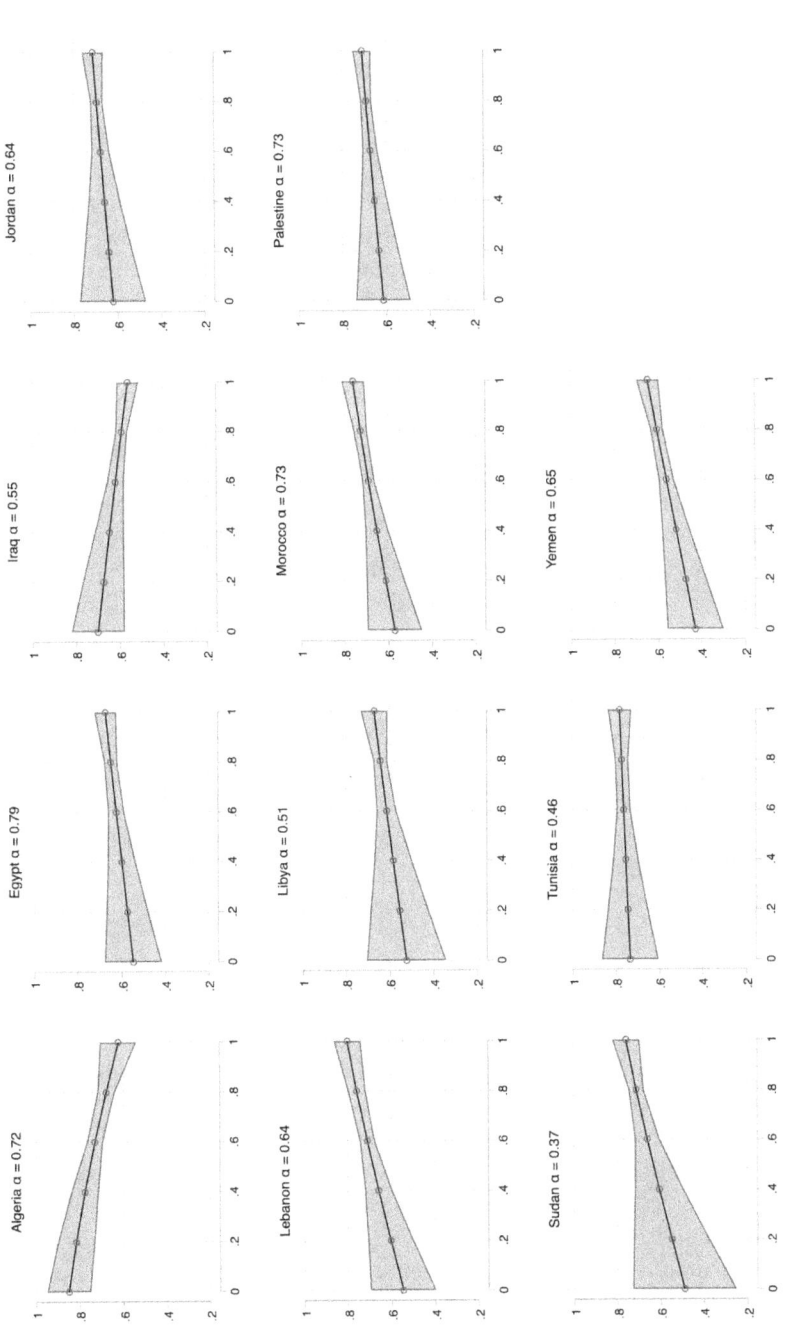

**FIGURE 4.2** Predicted probability of reporting climate change is a serious problem as a function of religiosity (country-specific Chronbach's alpha for the religiosity scale in parentheses)

change was a serious problem (whether "very" or "somewhat") as a function of respondents' religiosity. Additionally, each plot notes Chronbach's alpha (a statistic used to indicate the reliability of a scale) for the index in each country. Typically, for a four-item scale, a minimally acceptable alpha would be around .65 (DeVellis 2017, 136), yet the reliability of our key predictor is well below this threshold in a number of countries. That is to say, while theoretically the content that makes up the religiosity scale is clearly valid on its face, the items do not empirically hold together well in certain settings. The reasons for this statistical circumstance may vary—it could be anything from cultural idiosyncrasies to limitations in the polling infrastructure undermining the validity of the individual items and therefore the reliability of the combined index. In any case, the discussion will focus only on those countries for which the religiosity index is statistically informative.

Based on the results in Figure 4.2, the association between religiosity and attitudes toward climate change appears to be context dependent. Lebanon, Morocco, and Yemen all evidence a statistically significant positive correlation: as religiosity increases, so too does the probability of believing that climate change is a serious problem. The same tendency is apparent in Egypt, Jordan, and Palestine; however, statistically speaking, we cannot dismiss the possibility of these results being the product of random chance rather than reflecting the reality on the ground. Last, Algeria is the lone country where one's religious observance is negatively correlated with the gravity they ascribe to climate change. Overall, it seems like rather than asking *whether* religiosity is positively associated with climate change, a better question is *under what conditions* do we observe this relationship. This is but one of the potential avenues for future research on how Islam does (and does not) influence reactions to climate change.

## Conclusion

This chapter surveyed the role of Islam in shaping attitudes and behaviors around climate change. Doctrinally, the religion's primary texts and secondary interpretations offer as close to a consensus as there is for a tradition this long-standing and diverse. Of course, when it comes to translating doctrine to praxis, this near unanimity gives way to a variety of behaviors. At the elite level, political leaders of Muslim-majority states are often nested in structural contexts that directly pit commitments to climate mitigation and adaptation against their country's (and often their own) economic interests. Even when these leaders do advocate for environmental sustainability, their motivations are often untethered to any religious considerations. A number of NGOs at the national and international level, on the other hand, explicitly root their environmental activism in Islamic teachings, though their scope and efficacy are very much context dependent.

Last, mass-level public opinion evidences that majorities in select Arabic-speaking countries are concerned about climate change, though perhaps not at the level that one would expect given how acutely the negative by-products of a warming world will affect this region. Moreover, religiosity does not appear to uniformly correlate with these attitudes, although the mixed results may in part be due to country-level variation on what measures best capture "religiosity" as a construct. Given that the first climate-related questions asked in the ArabBarometer, the MENA region's premiere cross-national survey, appeared in the 2018–2019 wave, there is clearly still much to be learned about climate attitudes across this concentration of Muslim-majority states.

Indeed, each of these facets offers fertile ground for future research. At the macro level, it would be helpful to understand what benefits leaders accrue from climate diplomacy and what factors motivate or deter political elites from pursuing those benefits. In terms of civil society, more data on how (or whether) mosques are consciously building spacing and programs with sustainability in mind (largely, though perhaps not exclusivity in the West) would provide a fuller sense of the priorities of these communities' leaders and whether they align with those of their congregants.

Along those lines, efforts to gauge the hierarchy of policy preferences among Muslim populations should continue, though it would be advisable to move away from open-ended and rank-order questions and utilize methods that would provide more accurate and holistic data. For example, conjoint experimentation could be one way to unobtrusively but validly chart the policy landscape for these populations. Yet, even when we can reliably gauge Muslims' climate attitudes, we should be mindful that these are but snapshots. Brooke et al. (2023) show that elite messaging, particularly in religious settings, can meaningfully shift the stances Muslims take on important social issues, and there is no reason to believe that climate change constitutes an exception to this finding. Clearly, even though it may be evident what Islam has to say about climate change, our understanding of how this message is received and actualized will continue to evolve.

## Bibliography

Abnett, Kate. 2023. "UAE Says It's Committed to Meet CO2 Emissions Targets after Criticism." *Reuters*. https://www.reuters.com/world/middle-east/uae-says-committed-meet-co2-emissions-targets-after-criticism-2023-07-24/ (December 5, 2023).

Aman, Ayah. 2022. "First Climate-Related Fatwa Prohibiting Environmentally Harmful Practices Issued in Egypt." https://www.al-monitor.com/originals/2022/10/first-climate-related-fatwa-prohibiting-environmentally-harmful-practices-issued (December 7, 2023).

Arab Barometer. 2019. *Arab Barometer Wave V Questionnaire*. https://www.arabbarometer.org/wp-content/uploads/ABV_SourceQuestionnaire_ENG_Public-Opinion-2018-2019.pdf (December 7, 2023).

Arab Barometer. 2022. *Arab Barometer Wave VII Questionnaire*. https://www.arabbarometer.org/wp-content/uploads/ENG-Arab-Barometer-Wave-VII-Questionnaire-RELEASE-v2.pdf (December 7, 2023).

Brooke, Steven, Youssef Chouhoud, and Michael Hoffman. 2023. "The Friday Effect: How Communal Religious Practice Heightens Exclusionary Attitudes." *British Journal of Political Science*. 53(1): 122–139.

Brown, Nathan J. 2017. "Official Islam in the Arab World: The Contest for Religious Authority." *Carnegie Endowment for International Peace*. https://carnegieendowment.org/2017/05/11/official-islam-in-arab-world-contest-for-religious-authority-pub-69929 (December 6, 2023).

*"Caring for the Environment Like Prophet Muhammad (p)" Khutbah by Amira Al-Sarraf (4/20/18)*. 2020. https://www.youtube.com/watch?v=2SKM8290ioE (December 9, 2023).

Climate Action Tracker. 2023a. "Countries." *Climate Action Tracker*. https://climateactiontracker.org/countries/ (December 5, 2023).

Climate Action Tracker. 2023b. "Nigeria." *The Climate Action Tracker*. https://climateactiontracker.org/countries/nigeria/(December 6, 2023).

Climate for Peace. 2023. "North West Climate-Peace Hubs: A Climate Security Approach to Conflict Prevention | Climate-Diplomacy." *Climate Diplomacy*. https://climate-diplomacy.org/north-west-climate-peace-hubs-climate-security-approach-conflict-prevention (December 6, 2023).

DeVellis, Robert. 2017. *Scale Development: Theory and Applications* (4th ed.). London, UK: Sage Publications.

Drake, Bruce. 2013. "U.S., Middle East Publics Less Concerned about Climate Change than Those in Other Nations." *Pew Research Center*. https://www.pewresearch.org/short-reads/2013/11/11/u-s-middle-east-less-concerned-about-climate-change-than-those-in-other-nations/ (December 8, 2023).

Elgendy, Karim. 2021. "On the Bandwagon to Glasgow: Climate Action in the MENA Region." *Al Jazeera*. https://www.aljazeera.com/opinions/2021/10/29/on-the-bandwagon-to-glasgow-climate-action-in-the-mena-region (December 5, 2023).

Energy Institute. 2023. *Statistical Review of World Energy*. London, UK.

"Enviromuslims." 2023. https://enviromuslims.ca/ (December 9, 2023).

Falk, Thomas O. 2021. "Morocco: MENA's Rare Climate Change Success Story." *Al Jazeera*. https://www.aljazeera.com/news/2021/11/10/morocco-leads-the-fight-against-climate-change-in-the-middle-east (December 6, 2023).

Friedlingstein, Pierre et al. 2023. "Global Carbon Budget 2023." *Earth System Science Data*. 15(12): 5301–5369.

"Great Big Green Week." 2023. *Muslim Charities Forum*. https://www.muslimcharitiesforum.org.uk/2023/06/05/great-big-green-week-203/ (December 9, 2023).

"Green Muslims." 2023. *Green Muslims*. https://www.greenmuslims.org (December 9, 2023).

"Greening Canadian Mosques." 2023. *Faith & the Common Good*. https://www.faithcommongood.org/gcm (December 9, 2023).

*From Symbolic to Systematic in Response to Climate Change*. 2020. https://www.youtube.com/watch?v=4_ZceTWHg48 (December 6, 2023).

"IFEES." 2023. *The Islamic Foundation for Ecology and Environmental Sciences (IFEES)*. https://www.ifees.org.uk/ (December 9, 2023).

Islamic Declaration on Global Climate Change. 2015. https://www.ifees.org.uk/wp-content/uploads/2020/01/climate_declarationmmwb.pdf (December 9, 2023).

Islamic Relief Worldwide. 2020. "Adapting for Climate Justice." London, UK.

Izzi Dien, Mawil. 1992. "Islamic Ethics and the Environment." In *Islam and Ecology*, edited by Fazlun Khalid and Joanne O'Brien. London: Cassell Publishers Ltd, 25–35.

Izzi Dien, Mawil. 2000. *Environmental Dimensions of Islam*. Cambridge: Lutterworth Press.
*Jumu'ah Khutbah | English | Islam and the Environment*. 2021. https://www.youtube.com/watch?v=pdnb_qK27xI (December 9, 2023).
Keynoush, Banafsheh. 2023. "Climate-Induced Migration in the GCC States: A Looming Challenge." *Middle East Institute*. https://www.mei.edu/publications/climate-induced-migration-gcc-states-looming-challenge (December 5, 2023).
Khalid, Fazlun. 2002. "Islam and the Environment." In *Encyclopedia of Global Environmental Change: Social and Economic Dimensions of Global Environmental Change*, edited by Peter Timmerman and Ted Munn. Chichester, United Kingdom: John Wiley & Sons, Ltd, 332–339.
Khalid, Fazlun. 2019. *Signs on the Earth: Islam, Modernity and the Climate Crisis*. Markfield, Leicestershire, England: Kube Publishing Ltd.
Leiserowitz, A. et al. 2021. *International Public Opinion on Climate Change*. New Haven, CT: Yale Program on Climate Change Communication and Facebook Data for Good.
Nasr, Seyyed Hossein. 1996. *Religion and the Order of Nature*. New York: Oxford University Press.
Nasr, Seyyed Hossein.. 1998. "Sacred Science and the Environmental Crisis: An Islamic Perspective." In *Islam and the Environment*, edited by Haleem, Harfiyah A. London: Ta-Ha Publishers Ltd, 118–137.
Ozdemir, Ibrahim. 2022. "The Concept of Al-Mizan (Balance) as a Framework for a New Ethics of Environment and Sustainability." In *Creation - Transformation - Theology: International Congress of the European Society for Catholic Theology*, edited by Margit Eckholt. Berlin Münster: LIT Verlag.
*Quran Inspires Environmental Stewardship - Sr. Emmalee Aman*. 2015. https://www.youtube.com/watch?v=bLTIkzNgrJ8&ab_channel=ICNA (December 6, 2023).
Schlosberg, David, and Lisette B. Collins. 2014. "From Environmental to Climate Justice: Climate Change and the Discourse of Environmental Justice." *WIREs Climate Change*. 5(3): 359–374.
State Information Service. 2022. "President El-Sisi's Speech at the Opening Session of COP 27." https://www.sis.gov.eg/Story/172566/President-El-Sisi%27s-Speech-at-the-Opening-Session-of-COP-27?lang=en-us (December 5, 2023).
Vohra, Anchal. 2023. "The Middle East Is Becoming Literally Uninhabitable." *Foreign Policy*. https://foreignpolicy.com/2021/08/24/the-middle-east-is-becoming-literally-uninhabitable/ (December 5, 2023).

# 5

# COMPASSION AND INTERDEPENDENCE IN THE AGE OF CHANGING CLIMATES

Buddhist Understandings of Human-Environmental Relationships in the Anthropocene

*Kalzang Dorjee Bhutia and Amy Holmes-Tagchungdarpa*

### Introduction

In 2015, a group of Buddhists presented a joint statement, signed by Buddhist leaders and scholars from around the world, at COP 21, the United Nations Climate Change Conference. In the statement, the authors, who included David Loy (b. 1947, an American Buddhist scholar), Ven Bhikkhu Bodhi (b. 1944, an American Buddhist monastic), and John Stanley (b. 1947, an American biologist), outlined the impact of climate change around the world and supported the call by scientists to reduce carbon emissions to "no more than 350 parts per million" (Loy et al. 2015). Signatories included political leaders, Buddhist teachers, and scholars from the United States, "Bangladesh, Japan, Korea, Malaysia, Mongolia, Myanmar, Sri Lanka and Vietnam" (OneEarth Sangha 2015). In the years that followed, people continued to sign on, and OneEarthSangha, a website that promoted the statement, estimated that there now may be "millions" of signatories. This statement was a significant symbol of unity between very diverse Buddhist communities. To be clear, despite the common origin of Buddhism in South Asia 2,500 years ago, Buddhism has traveled around the world, and there is now no central teaching, common community, or institution that has authority over what Buddhism is or what it means to be Buddhist. Initiatives like the 2015 statement are therefore unusual.

What allowed for so many different religious, cultural, and political powers to work in unity? At the outset, the statement appears to largely summarize well-established scientific perspectives on climate change. However, as it

DOI: 10.4324/b23076-5

continues, the authors incorporate specifically Buddhist discourse. For example, this section contains many Buddhist concepts:

> The four noble truths provide a framework for diagnosing our current situation and formulating appropriate guidelines—because the threats and disasters we face ultimately stem from the human mind, and therefore require profound changes within our minds. If personal suffering stems from craving and ignorance—from the three poisons of greed, ill will, and delusion—the same applies to the suffering that afflicts us on a collective scale. Our ecological emergency is a larger version of the perennial human predicament. Both as individuals and as a species, we suffer from a sense of self that feels disconnected not only from other people but from the Earth itself.
> *(Loy et al. 2015)*

These include the invocation of the four noble truths (that life is suffering; that this suffering comes about through attachment; that there is a path out of suffering; and that the path is the eightfold path, which prescribes ethical and contemplative guidance); the three poisons; and attachment to the self. These ideas unite different Buddhist traditions around the world and across time, even if their definitions and practices related to them can vary considerably.

The release of the statement in 2015 was part of the development of a broader popular cultural movement that associated Buddhism with the environment. The development of this association has not been accidental but instead has come about due to a variety of trajectories, including the rise of the environmental movement in the 1970s around the world, which corresponded with the popularization of Buddhism as a countercultural religion/philosophy (Elverskog 2020). In recent years, scholars including historian Johan Elverskog have demonstrated the danger of this stereotyping, particularly due to concerns that promoting this vision of a Green Buddhism flattens the diversity and complexity of Buddhist histories and communities (Elverskog 2020).

Despite this important warning, books, organizations, and events continue to promote this message. So do Buddhist authorities. In the last decade, two of the most well-known Buddhist teachers in the world have written and spoken extensively on climate change. The late Vietnamese Buddhist teacher Thich Nhat Hanh (1926–2022) warned that:

> [t]the truth is that many of us have become alienated from the Earth. We forget that we are alive, here on a beautiful planet and that our body is a wonder given to us by the Earth and the whole cosmos.... Only when you have this right view, this insight, will discrimination no longer be there,

and there will be a deep communion, deep communication between you and the Earth. All kinds of good things will come from it. You transcend the dualistic way of seeing things: the idea that the Earth is only the environment, and that you are in the center; and that you only want to do something for the Earth so *you* can survive.

*(Thich Nhat Hanh 2022, 2, italics from original)*

Here, Thich Nhat Hanh invokes the relatedness between humans and the environment as a way to encourage people, particularly young people, to take action, or, in his words, to "wake up" (Thich Nhat Hanh 2022, 4). His colleague and occasional conversation partner, Tibetan Buddhist leader the Fourteenth Dalai Lama Tenzin Gyatso (b. 1935), also released a co-authored book where he urged the younger generation to "be rebels demanding climate protection and climate justice because your future is at stake" (Dalai Lama 2022 94). He argued that "[w]e need a revolution of compassion based on warmheartedness that will contribute to a more compassion world with a sense of oneness of humanity. The entire human family must unite and cooperate to protect our common home" (Dalai Lama 2022, 97).

The resonance between both messages is striking, though not surprising. Both of these teachers invoke Buddhist concepts, including interconnectedness and compassion, to motivate action. The concepts of interconnectedness (sometimes also called interdependent origination) and compassion have been invoked by Buddhist environmentalists for decades. The fact that initiatives such as these books and the 2015 statement continue to appear indicates that Buddhist ideas can be used successfully in mobilizing awareness of, and action in response to, climate change. In this chapter, we will provide evidence of this argument but also advance that these initiatives are by no means hegemonic or straightforward. Instead, Buddhist attitudes toward climate change, and the environment more generally, are diverse and complicated and should be understood in longer histories and specific contexts. We will demonstrate this through analysis of scholarly and popular literature on Buddhism and climate change and incorporate our own literary and ethnographic research. Kalzang Dorjee Bhutia is a scholar of Buddhism, the environment, and Indigenous Himalayan culture who has been educated in both traditional Himalayan Buddhist and secular university environments and draws on lifelong experience to think critically about how communities in his home state of Sikkim and surrounding regions relate to their environment. Amy Holmes-Tagchungdarpa studies literary, material, and visual cultures of the Himalayas, from India to China, in order to understand environmental change and its impact on local communities, and has been trained in Asian Studies, history, and anthropology in secular universities. We bring these analytical perspectives and biases to how we engage with this material.

In our survey of sources, we have found Buddhism and climate change are talked about using three approaches: the first is that scholars tend to explore Buddhist philosophy and concepts as models for responding to environmental change. Much of this scholarship is inspired by questions that can be hypothetical – what would Buddhist thinkers from the past as well as present do to respond to the impact of climate change, and can these ways of thinking be useful for developing responses? The second is the study of how the environment is engaged with specific historical or living Buddhist communities. Some of this scholarship does not attend to climate change but is helpful and important for gaining insight into what Buddhists may contribute to broader discussions. Sometimes these two ways of talking about Buddhism and climate change come together; for example, in discussions about Engaged Buddhism, which emphasizes social justice.

However, these different emphases are not always attentive to what Buddhists are actually doing and thinking. In this chapter, we will outline the three approaches and consider the ways that they can intersect, as well as how communities use different forms of knowledge to respond to climate change and work for alternative futures. We will consider philosophies and cosmologies related to climate change and the environment that have been used to mobilize effective activism but also remain attentive to how ritual and local forms of knowledge provide important insights that should be centered in creating inclusive forms of action.

## Buddhist Philosophy as a Resource for Responding to Climate Change

The 2015 statement signed by Buddhist authorities was authored by three American Buddhists. The statement represents a modernist construction of Buddhism that has been traced by Buddhist studies scholar David McMahan. This construction centers on the compatibility of Buddhism with science and empirical rationalism and emerged from colonial engagements between Asians, Europeans, and Americans (McMahan 2008). This modernist, environmentally friendly Buddhism has fueled the publication of much work on how Buddhists think about the environment, including the edited volumes *Dharma Rain* (Kaza and Kraft 2000) and *Dharma Gaia* (Hunt Badiner 1990) that drew together primary sources related to Buddhism and the environment and scholarship. This scholarship was often inspired by historical Buddhist conceptions of the environment but specifically sought to consider how these historical conceptions could be used in the present day.

These specific readings of Buddhist sources have led to the development of practice communities and programs. Well-known teachers in these movements include environmental activist Joanna Macy (b. 1929) and poet Gary Snyder (b. 1930). Macy studied with Buddhist teachers in the Himalayas,

India, Sri Lanka, and Thailand in the 1960s and went on to write prolifically and develop workshops around how to understand interdependence, or dependent co-arising, as a way to focus environmental activism (Macy 1991). Along with his work as an activist, Snyder famously used the Buddhist textual form of the sutra to compose "The Smokey the Bear Sūtra," which stars Smokey the Bear, the icon of the U.S. Forest Service, acting as a Bodhisattva to thwart "capitalism and totalitarianism" and understand the truth, that "all is contained vast and free in the Blue Sky and Green Earth of One Mind" (Sydney 1969). The study of scholar-activists such as Macy and Snyder is widespread in contemporary Buddhist communities in the United States and is often undertaken with a specific goal in mind: the use of Buddhist ideas to deal with the environmental impact of climate change. In an article that represents this trend, religious studies scholar Charles R. Strain examines how practices they have developed, along with tree ordination (which will be discussed later in this chapter), "might evolve towards greater efficacy" (Strain 2016, 140) to "meet the challenge of climate change" (Strain 2016, 138). Strain argues that these practices encourage people to move beyond abstract approaches to climate change by taking concrete action and that Buddhism provides a "moral strength" to do so. (Strain 2016, 153) He argues that "Buddhist communities ... could join transnational advocacy networks to provide needed support to indigenous resisters" (Strain 2016, 153). This argument is important as an example of how people talk about Buddhism and climate change in a prescriptive way, encouraging specific types of action.

David Loy, one of the co-writers of the 2015 statement, has also written extensively on Buddhism and climate change. In a recent book, *Ecodharma*, Loy outlines the challenges of climate change and discusses the "potential" for Buddhism to respond to climate change. He argues that this potential needs to be realized since "some traditional Buddhist teachings discourage us from social and ecological engagement" and emphasize the "*otherworldly* orientation" of Buddhism (Loy 2019, 13). Loy writes to encourage an audience of "us" to undertake "the *ecosattva* path." This path brings together ecological concerns with the ideal of the compassionate bodhisattva, which Loy argues is necessary as "many Buddhists... have usually been taught to focus on their own peace of mind" (Loy 2019, 233).

Both Strain and Loy emphasize the significance and potential for Buddhism as a path of action. However, their readings of Macy, Snyder, and other Buddhist thinkers are based on an assumption of nonaction and an audience that sees Buddhist practice as separate from other activities. This is unfortunate, as it does not take into account the very many ways Buddhists historically have acted and been socially engaged. Loy's work in particular does not take into account the relatedness and community at the center of Buddhist practice across the world for centuries and instead focuses on Buddhists who "meditate inside buildings with screened windows" (Loy 2019, 15) While his book has been inspiring for many, the "us" it addresses is apparently not most of the world's Buddhists.

## Buddhist Understandings of the Environment in Practice

An emphasis on Buddhist morality and ethics is common in popular American writing on Buddhism. However, historical and contemporary Buddhist communities around the world have never been as disengaged as the writers suggested (nor, it seems, as Macy and Snyder suggested). Practicing Buddhist communities have not understood their practice, and specifically, their relationships with their environments, as operating in a vacuum. Instead, the many impacts of climate change are often understood within broader cosmological frameworks in Buddhist communities that outline how humans relate to the world around them. Although Ian Harris (1991) and other scholars have argued that early Buddhism did not demonstrate an interest in the environment, alternative interpretations of historical materials and ethnographic research illustrated the many complicated ways that Buddhism has been deeply connected to local ecologies. Examples of ritual traditions in Buddhist societies illuminate the cosmological connections between humans, nonhuman animals, and other more-than-human forces and agents in the environment.

### *May All Beings Be Free from Suffering: Life Release as Human-Nonhuman Animal Ethics*

One vivid example is the practice of life release (Chinese: *fangsheng*; Japanese: *hōjō-e*; Tibetan: *tshe thar*), which is practiced in many East and Inner Asian Buddhist communities and also in non-Buddhist communities in parts of Asia (Shiu and Stokes 2008, 185). Life release itself is not a specific ritual but instead is a type of ritual where nonhuman animals who were otherwise intended for slaughter or a life in captivity undergo a ritual transformation by being freed by ritual practitioners to live out their lifespans without human intervention. A simple explanation of the motivation for this ritual is that it accrues merit, or karma, for the people who sponsor the life release as it is a compassionate act. For this reason, it could be argued that life release represents the anthropocentric elements of the Buddhist tradition, whereby care for nonhuman animals is predicated on the benefits humans will accrue for ritual action. Scholars have attempted to trace its history and argued that it demonstrates a manifestation of the ancient Indian Buddhist concept of *ahimsā*, nonviolence (Shiu and Stokes 2008, 183).

This would support specific interpretations of Buddhist texts and cosmologies that also appear to center human interests. For example, in the popular Tibetan Buddhist cosmological framework that posits there are six realms of existence, animals appear below humans. In the famous nineteenth-century Tibetan language text *Words of My Perfect Teacher* (Classical Tibetan: *Kun bzang bla ma'i zhal lung*), the Buddhist teacher Patrul Rinpoche (1808–1887) explained why this was the case:

> To practice the real, authentic Dharma, it is absolutely necessary to be a human being. Now, suppose that you did not have the support of a human form, but had the highest form of life in the three lower realms, that of an animal – even the most beautiful and highly prized animal known to man. If someone said to you, "Say Om mani padme hum once, and you will become a Buddha," you would be quite incapable of understanding his words or grasping their meaning, not would you be able to utter a word. In fact, even if you were dying of cold, you would be unable to think of anything to do but lie in a heap – whereas a man, no matter how weak, would know how to shelter in a cave or under a tree, and would find wood and make a fire to warm his face and hands. If animals are incapable of even such simple things, how would they ever conceive of practicing Dharma?
>
> (Rinpoche 1988, 22)

Nonhuman animals are therefore represented as inferior to humans because of their mental faculties. However, later in the text, Patrul Rinpoche complicated this representation. He discussed how humans actually have very little compassion, as they mistreat nonhuman animals for their own gain. Patrul used the illustration of how yaks are treated as beasts of burden and how humans, for all their mental faculties, become angry and cruel and feel "not the slightest compassion" for the animal who works on their behalf (Patrul 1988, 204). This treatment contributes toward the position of animals in the realms of existence. Patrul's argument here is a critique of human cruelty that keeps other beings below them.

Since, according to Patrul Rinpoche and other Buddhist authorities, at some point all beings were our mothers, as all the sentient beings in the realms of existence are part of the same cycle of rebirth, kindness, and compassion toward nonhuman animals are important. Within these broader traditions, freeing life takes on additional layers of meaning. In her deep ethnography of pastoralists living in Eastern Tibetan regions of contemporary China, anthropologist Gillian Tan discusses the close relationships between pastoralists and their yaks. She provides vivid examples of the care pastoralists exhibit toward their yak, which is known locally as *"nor"* or wealth, as yaks provide

> the basis for life on the grasslands: through their milk from which yoghurt, butter and cheese are made, their dung from which fuel for burning comes,

their hair and fur from which the tent, slingshots, and ropes are woven, and eventually their meat, which is eaten, and skin, which is used to make bags and rugs.

*(Tan 2016, 4)*

While this conception of wealth as merely economic wealth may further demonstrate the ways Buddhists demonstrate anthropocentric attitudes towards the natural world, studying freeing life in this context demonstrates how wealth is instead seen here as more multi-faceted. This is because freeing life as an offering (Tibetan: *mchod*) or as a gift (Tibetan: *sbyin*) to deities "augments the wellbeing of the household" but also "the fortunes of the overall herd" (Tan 2016, 4). Here then, the specific yak who has been freed benefits, but so does the rest of the herd, and these fortunes are not only economic but also related to broader conceptions of health and wellbeing. The well-being of humans and nonhuman yaks here is interconnected, and the treatment and care for the yaks exhibited by pastoralists confirms this. Freeing life rituals on the Tibetan plateau and in surrounding regions of Inner Asia and the Himalayas will often entail a blessing from a lama or Buddhist practitioner, and then the nonhuman animal (including yaks, but also cows, chickens, goats, and other nonhuman animals) is not tied or tagged, is free to graze, and is not used to carry loads. Yaks no longer have their hair and fur cut. Most importantly, freed nonhuman animals cannot be killed, though their bodies may still be consumed once they die from natural causes (Tan 2016, 5).

In China, the exact origin of the practice of freeing life is unclear, but an imperial decree from 759 CE records that the dynasty at the time sponsored the construction of eighty-one ponds for the "release and protection of fish" (Stevenson 1998, 394). This suggests that this practice was already known, and other public and private ceremonies to release fish, birds, turtles, and other beings were undertaken throughout the imperial period and continue to be popular today in Chinese communities around the world. In China, these rituals are presided over by Buddhist monastics, and in the imperial period, they often followed a set formula, described here by Buddhist studies scholar Daniel Stevenson:

The officiating priest begins by asking the three jewels [the Buddha, the Dharma, and the Sangha] to purify the assembled creatures of the mental defilements that prevent them from comprehending what is about to be said. The priest than bestows refuge in the three jewels and recites the epithets of Ratnabhava so that the animals will be reborn in their next life in the Heaven of the Thirty-Three and, eventually achieve enlightenment. This is followed by a lecture on the twelvefold chain of dependent origination, one of the more difficult of Buddhist doctrines even for humans, then a confession of the animals' sins, and finally a prayer for rebirth in the pure land.

*(Stevenson 1998, 395; italics removed from original and square brackets added by authors)*

This ritual is noteworthy, as it demonstrates that nonhuman animals have the same inherent capacity for enlightenment as humans, which again troubles the idea that nonhuman animals are inherently inferior and claims to the anthropocentrism of Buddhist traditions.

Today, Buddhist communities around the world continue to free life, but these ritual traditions have become controversial. There are widespread ethical arguments about the breeding of specific species, especially birds and turtles, specifically for the purpose of animal release, and especially concerns that in some places, these animals are released and then recaptured to maximize profit. Scholars Henry Shiu and Leah Stokes have argued that life release can now have unforeseen negative consequences for local environments, especially as invasive, nonnative species are released into ecosystems, bringing new environmental issues as a result (Shiu and Stokes 2008, 190–192). These environmental issues can exacerbate those already connected to climate change. However, despite these issues, life release remains popular and demonstrates concerns about and recognition of the close karmic connections between humans and nonhumans in Buddhist communities.

### *Ordaining Trees: Humans and the Forest in Northern Thai Buddhism*

Buddhist ritual life also includes plants. In Thailand, the novel tradition of tree ordination has emerged since 1988 in response to deforestation, undertaken to promote the exports of cash crop agriculture (Strain 2016, 140). Anthropologist Susan Darlington has studied tree ordinations in Thailand for decades and has demonstrated the complex and multifaceted nature of these rituals as a way to bring attention to the environmental problems associated with deforestation and other environmental issues, but also has a way to understand relationships between Buddhist monastics, the human lay communities who support them, and the forests in which both monastics and laypeople live.

> The rituals and the trees wrapped in orange robes remind villagers of their dependence on the forest for their livelihoods—food, materials for daily life, and water. As monks depend on the laity for their material needs, so too the forest depends on the people who live around it for preservation. People can either protect the forest or cut it down... The movement is not about trees per se, but the monks and the people with

whom they live and work who must deal with the direct consequences of environmental destruction.

*(Darlington 2013, 4)*

This context is important because ordination across Buddhist communities is a deeply significant ritual that brings about transformations that allow humans to act as carriers and, in some instances, manifestations of the Buddha's teaching. By undertaking ordination and vowing to follow the *Vinaya*, or monastic law it entails, Buddhist monastics gain tremendous moral authority in their communities. What does it mean to ordain a tree then – can a tree carry this authority? Darlington demonstrates that this is not simple. Even Phrakhru Manas Natheephitak, the first monk to carry out a tree ordination, "did not intend to ordain a tree. He performed a ritual to consecrate a forest and seedlings for reforestation to raise awareness of people's dependence on them and to object to deforestation occurring due to logging" (Darlington 2013, 4). It was villagers who started to use the term "ordained trees" (Thai: *ton mai buat*), and this association has had great resonance because of its invocation of the authority of ordained monastics in Theravada Thai society. The monastics who continued to engage in this movement, known as "environmental monks" (Thai: *phra nak anaraksa thamachat*), were part of broader movements that aimed to alleviate the suffering of lay people and "challenge the power of political and business interests" that were fueled by "greed, anger, and ignorance (the root evils in Buddhist teachings)" (Darlington 2013, 6).

The efficacy of tree ordination as a form of activism was undoubtedly due to the power of Buddhist monastics. However, just as important was the worldview that monastics and laypeople hold of the forest, which is not just a site for resource extraction. For northern Thai monastics and laypeople, the forest is also home to deities, spirits, and other unseen forces who need to be acknowledged in order for villagers to continue to live with the forest. Susan Darlington traces how influential environmental monk Phrakhru Pitak supported the villagers in his village, "requesting the permission and support of the village guardian spirit" before ordaining trees. After the ordination, the forest was considered consecrated, and villagers were careful to follow regulations for forest care that they established with Phrakhru Pitak in case they invited "the spirit's wrath if they violated the community forest regulations" (Darlington 2007, 174). The fear of the power of the guardian spirit encouraged villagers to care for the forest, and ritual communications between the spirits of the forest, monastics, village spirit specialists, and villagers promoted respect and, as Darlington observed, had material benefits as the forest remained healthy, which was understood as a sign that appropriate conduct was followed by humans (Darlington 2007, 174). This incorporation of local beliefs with Buddhist concepts illustrates the capacious nature of Buddhism to adapt to local environments in the practice of the Buddhist concept of "skilful means" and respond to different challenges and contexts.

However, local beliefs do not always travel with ordination rituals. The efficacy of tree ordination has also led to the appropriation of tree ordination by the state and mainstream groups. As Darlington argued,

> [t]he contexts for the rituals have changed. Rather than pushing people to question modern, consumerist values as causes of environmental destruction and human suffering, they are increasingly used to support national agendas and to undermine the power of the rural people whom environmental monks aim to help.
>
> *(Darlington 2007, 171)*

But even as the impact of these rituals is challenged through widespread adoption, the image of ordaining trees continues to draw attention to the important ways that Buddhism and the environment are related, especially how living Buddhist communities conceive of the world around them as agentive and in need of recognition and care.

## Buddhist Action in Response to Climate Change

These diverse conceptions of the environment and its interdimensional residents have inspired the emergence of social movements concerned with environmental justice, and specifically, climate change. The conception of Buddhism as a "green," or environmentally aware, religion is well-established in popular culture. As Johan Elverskog has argued, these popular cultural ideals do not necessarily represent the complexities of Buddhist attitudes toward the environment in historical contexts, but they have influenced new movements and activism within Buddhist communities (Elverskog 2020). Buddhist movements that promote awareness of climate change, and concern to respond to the many interconnected impacts of climate change in different global contexts, are now found across Buddhist communities. Importantly, many of these forms of activism emerge from earlier, historically situated forms of Buddhist practice, thereby demonstrating continuity rather than radical change and challenging depictions of Buddhism as environmentally apathetic or unengaged. Importantly, these examples move beyond philosophical analysis and provide concrete evidence of how Buddhist communities are engaging practically in efforts to promote an awareness of climate change and mitigate the impacts on communities around the world in some of the locations in the oceans and mountains that are already heavily impacted by climate change.

## Eating to Mitigate Climate Change: Vegetarian Movements in Taiwanese Buddhist Communities

A common stereotype about Buddhists is that all Buddhists are vegetarian. This idea has been generated by the early Buddhist idea of *ahimsā*, or nonviolence, extolled in Buddhist monastic settings, and by the ideal of the Bodhisattva, a being who aims to be reborn again and again to help all sentient beings escape the suffering of cyclical existence. Historically, this was only the case in specific Buddhist communities and contexts, particularly in China and Korea, and particularly among monastics or people vowing to become vegetarian for certain periods of time (Kieschnick 2005). In other Buddhist communities, Buddhists ate meat as part of their traditional foodways, but often with ethical expectations. Religious Studies scholar Geoffrey Barstow has written about the complex ways Buddhists in Tibet thought about meat eating, especially through the lenses of Buddhist ethics and local cultural expectations connected to health and wellbeing, and has demonstrated that there has never been a singular narrative (Barstow 2017).

In Taiwan, Buddhist-influenced vegetarianism is widespread and has been part of movements to promote awareness of climate change. The Tzu Chi Foundation was founded in 1966 by the nun Master Cheng Yen (b. 1937) and has been engaged since in a variety of humanitarian causes, including disaster relief and environmental education. Master Cheng Yen first began to promote "environmental protection" in 1990, and Tzu Chi has since become well-known for recycling initiatives in Taiwan and for promoting "reduced consumption" (Zimmerman-Liu 2019). Promoting vegetarianism is part of Tzu Chi's environmental education initiatives. Tzu Chi has used two specific programs in support of encouraging vegetarianism: Ethical Eating Days, where on specific days people pledged to "unite" to eat a vegetarian diet to "prevent global warming and combat climate change" (Ethical Eating Day 2022); and Very Veggie, where people pledge to abstain from eating meat for time periods between one day and their entire lives (Very Veggie 2023).

These programs are connected with other global initiatives that span beyond Buddhist communities. For example, one of the partners of Ethical Eating is Meatless Monday, a program started in the United States in 2003 that encourages people to "skip meat once a week" and thereby "combat climate change," among other goals (Meatless Monday 2023). These connections are not surprising, as even Tzu Chi's website does not emphasize Buddhist concepts to promote vegetarianism but instead features discussions of environmental impact. This emphasis on science is part of Tzu Chi's broader environmental messaging, which combines scientific information with "Buddhist ethics and practical lifestyle tips" to promote concern for, and preservation of, the environment (Zimmerman-Liu 2019, 253). Specifically, these ethics include the promotion of the Bodhisattva path – the

Mahayana ideal of working for the good of all sentient beings – as part of the aspiration of Tzu Chi followers. Rather than focusing on becoming a bodhisattva in future lives, Zimmerman-Liu has demonstrated how Master Cheng Yen's teachings promote becoming a bodhisattva "in society today" by developing "compassion and great love for all humans, for society, and all living things" (Zimmerman-Liu 2019, 64). Tzu Chi's approach combines Buddhist concepts with science and is therefore inclusive of multiple worldviews. This approach demonstrates how Buddhist perspectives can interact with science to promote climate change awareness.

### Dealing with Waste in a Sustainable Way: Movements against Plastic Waste in the Himalayas

In the Himalayas, the most significant and pressing evidence of climate change has been the melting of the glaciers in the mountains. In 2023, the International Centre for Integrated Mountain Development based in Kathmandu released a report that demonstrated how, based on current melt, glaciers in the Hindu Kush Himalaya are expected to lose 30–50% of their volume by the year 2100 (ICIMOD 2023, xiii). The melting of glaciers has led to floods downriver and many other challenges for Himalayan communities but also renewed concerns about tourism in the mountains and the waste that it produces that contributes to greenhouse gas emissions. In 2020, international newspapers reported on discoveries of microplastics in ice and ice melt on Jomolangma, the Indigenous name for Mount Everest. This discovery contributed to widespread discussions about waste disposal, especially for the many climbers who now ascend the mountain. For some Himalayan Buddhists, this discovery was not surprising but affirmed long-standing concerns about the cumulative effect of mass tourism.

Long before these reports emerged, Sherpa entrepreneur Ang Dolma Sherpa, the founder of Utpala Craft, had begun a project to eradicate plastics from Buddhist material culture in Nepal. One of the most common images found on Google searches and postcard images of Nepal is of prayer flags (known in many Himalayan languages as *lungta*), multi-colored cloth flags raised high in the mountains across passes. These flags are printed with Buddhist prayers for the benefit of all sentient beings. Historically, they were offered for many reasons in many places: to acknowledge power deities present in the mountains and valleys of the Himalayas, or as offerings to attract good forces and dispel negative ones. In Sherpa and other Himalayan communities, prayer flags were made of cotton and printed from woodblocks with ink made from plants. In the 1980s, new prayer flags made from cheaper synthetic fabrics and printed using ink made from plastics using silk screen technology began to circulate in China, Nepal, and India, and eventually, these new mass-produced prayer flags became more common than traditional flags. In the early 2010s, Ang Dolma Sherpa became concerned about how these flags were disposed of. Traditionally, prayer flags would break down by themselves in the elements, or if they remained, had to be burned to avoid the accumulation of ritual pollution (known in Tibetan and many other Himalayan languages as *drip*). Synthetic prayer flags do not break down over time as easily and have therefore contributed to the accumulation of microplastics in the glaciers of the Himalayas and if these flags are burned, they let off toxic fumes. In response to this concern, Ang Dolma Sherpa has designed biodegradable prayer flags made from cotton that incorporate traditional knowledge and encourages communities in Nepal to return to making their own prayer flags using traditional materials. While many have eagerly embraced this return, a challenge to this movement has been the comparatively higher cost of biodegradable materials. Ang Dolma Sherpa is eagerly seeking out alternatives and creating networks with other communities throughout the Himalayas to work out a cost-accessible model (Holmes-Tagchungdarpa 2023). This important alternative is part of broader discussions about the problems with plastic waste in the Himalayas and elsewhere across the Buddhist world.

### Rituals to Counter the Melting Glaciers: Traditional Knowledge in the Understanding of Climate Change

Both of these initiatives to counter climate change are connected to wider global movements and scientific discourses around climate change. However, they are also both deeply connected to Buddhist concepts and practices. To what extent can some of the Buddhist worldviews discussed earlier in the chapter also be understood as promoting action against climate change? Buddhist rituals such as ordaining trees and releasing animals may appear to function more as ways to promote awareness of the environment than have actual outcomes that benefit the environment. As Buddhist studies scholars Kalzang Dorjee Bhutia (one of the authors of this chapter) and Cathy Cantwell have argued when analyzing rituals that appear to be focused on the environment, a broader context is important. Many rituals in the Himalayas emphasize reciprocity between humans and spirits and force residents in specific places. But they do not necessarily always promote outcomes that can be understood as environmentally friendly in contemporary worldviews (Bhutia 2021a; Cantwell 2021).

However, rituals can still provide insight into relationality and care for the environment. Anthropologist Karine Gagné has argued that rituals, folk songs, and other local knowledge constitute essential ways of knowing the environment and are forms of care that connect humans with the beings and forces in the world around them among Buddhist communities in the state of Ladakh in the western Indian Himalayas (Gagné 2019). She has described how the *Skin jug* ritual was undertaken by human communities who live in Tingmosgang to "ensure that there would be sufficient water for crops." Since the local glacier guardian, or *zhidak*, is especially "stubborn," "monks, musicians, children, and laypeople" climb the mountain in the upper village and appeal to the *zhidak*, as well as to the glacier and lake as their father and mother, respectively. "Because there are sources of fresh water, life's most fundamental resource, the glacier and lake symbolize a father and a mother who take care of their children" (Gagné 2021, 195–196). This ritual affirmation of kinship between humans and parts of the land ensures the continued flow of water. Since climate change is understood as a manifestation of weakening ties and a "disregard for the deities," the care expressed in the ritual counteracts this disregard (Gagné 2021, 197).

In the eastern Indian Himalayan state of Sikkim, ritual is also important as an affirmation of relationships between humans and forces of the land. Sikkim's *zhidak* are also understood to be resident throughout the landscape and are communicated with through rituals. Kalzang Dorjee Bhutia recalls how his father Chewang Rinzing (1925–2020), a lama from the Lhopo community, discussed his concerns that the mountains turning black and green, due to the melting of the snow, represented broader systemic imbalances (Bhutia 2021b). In the final decades of his life, Chewang Rinzin carried out multiple rituals of care for the environment. While he had always performed these rituals, in the context of climate change, he specifically discussed erratic weather patterns as a manifestation of global changes.

In particular, he prescribed the performance of the *Nesol*, or in English, *The Propitiation Rite for the Sacred Habitat of the Valley of Rice*, a ritual discovered from the landscape of Sikkim that was believed to have been hidden there by the Himalayan cultural hero Guru Rinpoche in the eighth century CE and found again in the seventeenth century. *The Propitiation Rite for the Sacred Habitat of the Valley of Rice* requests for "the rain to fall on time" along with other signs of auspiciousness and peace and is carried out regularly throughout the annual Buddhist ritual calendar in Sikkim on the fifteenth and twenty-fifth days of the lunar calendar (Bhutia 2021a). This periodic renewal of ties promotes awareness of the land and, like the *Skin jug*, is especially concerned with mitigating the material impacts of climate change for Himalayan communities and landscapes. These rituals are not only performative but are also part of broader cosmologies that encourage an awareness of the connectedness between humans and the world around them and inspire motivation to action.

### Conclusion

This chapter has demonstrated that there is no singular Buddhist attitude towards climate change, just as there is no singular Buddhism. The 2015 statement at the COP21 was an important moment of unity for Buddhist communities, but there are hugely diverse approaches for how to attend to climate that reflect many elements of contemporary experience: the different ways that national governments where Buddhists live are responding to climate change; the different ways that diverse ecosystems where Buddhists live are being impacted by climate change; and the many different philosophical, ritual, and cultural forms of knowledge that compose Buddhism for different Buddhist communities. At present, Buddhists continue to talk about climate change in very different ways. Some emphasize individual contemplative practice and ethical action as effective paths to mitigate climate change. Others see the impacts of climate change as indicative of longer histories of human impacts on the environment and use rituals to inspire action and awareness. Others take Buddhist concepts such as the bodhisattva path to heart, seeing activism as an extension of their spiritual journey toward enlightenment. Just as Buddhist philosophical systems aim to break down dualism, Buddhist approaches to climate change are not either/or but instead embrace an array of experiences and ways that Buddhist humans draw upon to understand and use to relate with other dimensions, including nonhuman animals and deities and spirits of the land. These diverse pathways underpin powerful forms of action in response to the many impacts of climate change, in Buddhist parlance, for all sentient beings.

## Bibliography

Barstow, Geoffrey. 2017. *Food of Sinful Demons*. New York: Columbia University Press.

Bhutia, Kalzang Dorjee. 2021a. "Purifying Multispecies Relations in the Valley of Abundance: The *Riwo Sangchö* Ritual as Environmental History and Ethics in Sikkim," *MAVCOR*. 5(2) (2021): 1–26.

Bhutia, Kalzang Dorjee. 2021b. "Living with the Mountain: Mountain Propitiation Rituals as Pedagogy for Human-Environment Relations in Sikkim." *Journal of Buddhist Ethics*. 28(2021): 261–294.

Cantwell, Cathy. 2021. "Reflections on Ecological Ethics and the Tibetan Earth Ritual." *The Eastern Buddhist*. 33(1 (2001)): 113–119.

Dalai Lama, Tenzin Gyatso, and Franz Alt. 2020. *Our Only Home: A Climate Appeal to the World*. New York: Hanover Square Press.

Darlington, Susan. 2007. "The Good Buddha and the Fierce Spirits." *Contemporary Buddhism*. 8(2): 169–185.

Darlington, Susan. 2013. *Ordination of a Tree*. Albany: State University of New York Press.

Elverskog, Johan. 2020. *The Buddha's Footprint: An Environmental History of Asia*. Philadelphia: University of Pennsylvania Press.

Ethical Eating Day. 2022. "Tzu Chi: Ethical Eating Day." https://tzuchi.us/ethical-eating-day#:~:text=Ethical%20Eating%20means%20a%20lot,one%20day%20at%20a%20time.

Gagné, Karine. 2019. *Caring for Glaciers: Land, Animals, and Humanity in the Himalayas*. Seattle: University of Washington Press.

Gagné, Karine. 2021. "The Vanishing of Father White Glacier: Ritual Revival and Temporalities of Climate Change in the Himalayas." In *Understanding Climate Change through Religious Lifeworlds*, edited by David Haberman. Bloomington: Indiana University Press, 183–207.

Harris, Ian. 1991. "How Environmentalist is Buddhism?" *Religion.* 21(2): 101–113.
Holmes-Tagchungdarpa, Amy. 2023. "Preserving Offerings, Prolonging Merit." *Journal of Worldwide Waste.* 6(1): 1–11.
Hunt Badiner, Allan. 1990. *Dharma Gaia.* Berkeley: Parallax Press.
ICIMOD. 2023. *Water, Ice, Society, and Ecosystems in the Hindu Kush Himalaya.* Kathmandu: International Centre for Integrated Mountain Development.
Kaza, Stephanie, and Kenneth Kraft. 2000. *Dharma Rain.* Boulder, CO: Shambhala.
Kieschnick, John. 2005. "Buddhist Vegetarianism in China." In *Of Tripod and Palate,* edited by Roel Sterckx. New York: Palgrave Macmillan, 186–212.
Loy, David. 2019. *Ecodharma.* Boston, MA: Wisdom.
Loy, David, Bhikkhu Bodhi, and John Stanley. 2015. "The Time to Act is Now: A Buddhist Declaration on Climate Change." https://oneearthsangha.org/articles/buddhist-declaration-on-climate-change/
Macy, Joanna. 1991. *World as Lover, World as Self.* Berkeley: Parallax Press.
McMahan, David. 2008. *The Making of Buddhist Modernism.* New York: Oxford University Press.
Meatless Monday. 2023. "Meatless Monday." https://www.mondaycampaigns.org/meatless-monday
OneEarth Sangha. 2015. "The Time to Act is Now." https://oneearthsangha.org/articles/buddhist-declaration-on-climate-change/
Rinpoche, Patrul(Ed.). 1998. *Words of My Perfect Teacher.* Translated by Padmakara Translation Group. Walnut Creek, CA: Altamira Press.
Shiu, Henry, and Leah Stokes. 2008. "Buddhist Animal Release Practices: Historic, Environmental, Public Health and Economic Concerns." *Contemporary Buddhism.* 9(2): 181–96.
Stevenson, Daniel. 1998. "Freeing Birds and Fish from Bondage." In *Buddhist Scriptures,* edited by S. Lopez Donald. London: Penguin.
Strain, Charles R. 2016. "Reinventing Buddhist Practices to Meet the Challenge of Climate Change." *Contemporary Buddhism.* 17(1): 138–156.
Sydney, Gary. 1969. "Smokey the Bear Sūtra." https://sacred-texts.com/bud/bear.htm
Tan, Gillian. 2016. "'Life' and 'Freeing Life' (tshe thar) among Pastoralists of Kham: Intersecting Religion and the Environment." *Études mongoles et sibériennes, centrasiatiques et tibétaines* 47.
Thich Nhat Hanh. 2022. *Zen and the Art of Saving the Planet.* New York: HarperOne.
Very Veggie Movement. 2023. "Very Veggie Movement." https://veryveggiemovement.org/partners
Zimmerman-Liu, Teresa. 2019. "Humanistic Buddhism and Climate Change." PhD Dissertation, University of California San Diego.

# 6
# CONCERNED ABOUT CLIMATE
The Catholic Church, Environmental Stewardship, and the Challenge to Brazil's Bolsonaro

*Lan T. Chu*

## Introduction

The significance of the Amazon region's role in the maintenance of the global environment is undeniable. It houses the world's largest rainforest, river basin, and aquifer and produces between 6 and 9 percent of the world's total oxygen. With almost two-thirds of the Amazon and 28 of its 30 million inhabitants located in Brazil, "There is no way of detaching forest and country: most of the Amazon is in Brazil and most of Brazil is in the Amazon" (Casarões and Farias 2022, 55).

Because of Brazil's relationship to the Amazon, it often becomes the subject of international focus when discussing climate change. Greater international concern occurred following Brazil's presidential election of far-right politician Jair Bolsonaro in 2019. In his first year in office, the Amazon suffered major losses because of his deforestation policies, prompting an outcry from the international community. As a result of his policies at the start of his presidency, deforestation reached a 12-year high: Brazil's National Institute for Space Research reported that 3,769 square miles of the Amazon had been destroyed – an area about 12 times the size of New York City and nearly the size of Lebanon. The area lost is a 30 percent increase from the year before and the highest net loss since 2008 (Moriyama and Sandy 2019). With the fires unleashed under Bolsonaro's presidency, parts of the Amazon now emit more carbon dioxide than they absorb. Experts have indicated that if 20–25 percent of the Amazon were to be cleared because of deforestation, the Amazon basin would experience "dieback," i.e., turning into a savanna, which would no longer allow it to absorb the Earth's carbon dioxide. Given that 17 percent of the Amazon Forest has now been lost, domestic and international concern for the Amazon is warranted (Moriyama and Sandy 2019).

During his presidency, however, Bolsonaro has resorted to nationalist rhetoric to preempt international intervention in the Amazon. In his speech to the UN General Assembly on September 24, 2019, Bolsonaro argued how Brazil deserved to enjoy economic freedom, was subjected to the continued colonialist spirit of countries seeking economic gain, and did not support the preempting of sovereignty in "the name of an abstract 'global interest.'" He upheld that any discussion of the Amazon Rainforest must respect Brazil's sovereignty, and he condemned "the attempts at instrumentalizing the environmental matter or indigenous policies toward external political and economic interests, especially those disguised as good intentions" (Bolsonaro 2019).

This tension between the local, national, and global has been a central part of the discussion regarding the Amazon. Unlike states, international organizations, and secular organizations, the Catholic Church has a distinct approach on how to manage the domestic-international tensions. Being both an international organization and a part of localized civil society, the Catholic Church can intervene as a norm entrepreneur to promote an integral ecology that can help shape the narrative on climate change towards the norm of a sustainable practice of shared, global responsibility. Drawing from the Catholic Church's 2007 convening of the Fifth Episcopal Conference of Latin America and the Caribbean in Aparecida, Brazil, Pope Francis' 2015 encyclical "Laudato Si," and the October 2019 Synod for the Amazon, the Church's expression of an integral ecology will be brought to light. These efforts are examples of how the Church's norm entrepreneurship has challenged the international community's treatment of the environment and Brazilian president Jair Bolsonaro's deleterious, far-reaching deforestation policies.

### The Catholic Church as Norm Entrepreneur

A constructivist approach is best suited for highlighting the Church's work as a norm entrepreneur. Among international relations scholars, constructivism has been recognized as the framework most appropriate for understanding religion's influence on state and non-state actors. Constructivism highlights the normative processes that could potentially reframe our understanding of the approach to the global community, and such processes are initiated by norm entrepreneurs (Snyder 2011). According to constructivist scholars Martha Finnemore and Kathryn Sikkink, norm entrepreneurs are non-state entities motivated by empathy, altruism, and ideational commitments to initiate a norm life cycle (Finnemore and Sikkink 1998, 898). Norm entrepreneurs

> call attention to issues or even "create" issues by using language that names, interprets, and dramatizes them... The construction of cognitive

frames is an essential component of norm entrepreneurs' political strategies since, when they are successful, the new frames resonate with broader public understandings and are adopted as new ways of talking about and understanding issues. In constructing their frames, norm entrepreneurs face firmly embedded alternative norms and frames that create alternative perceptions of both appropriateness and interest.

*(Finnemore and Sikkink 1998, 893)*

In effect, the norm life cycle influences norm leaders (states) to adopt new norms, i.e., norm emergence.

Through imitation and socialization with other states, the idea is that the new norm will cascade and finally be internalized by other states in the international community (Finnemore and Sikkink 1998, 894–896). In this regard, the Church uses its doctrinal resources to reframe discussions on the environment and climate change, and its efforts parallel the norm entrepreneur framework. One of the key components articulated in Catholic Social Teaching is the concept of stewardship. In relation to the environment, stewardship impresses upon the faithful, moral, responsible caretaking of the earth, which has been divinely provided. Such care of the earth must be done in a way that benefits both current and future generations. Thus, with stewardship, one is "called to protect people and the planet, living our faith in relationship with all of God's creation. This environmental challenge has fundamental moral and ethical dimensions that cannot be ignored" (United States Conference of Catholic Social Bishops 2005). While we may have the technological ability, the political power, and the economic interest to develop the earth for selfish means, the Church advocates for a reframing of our understanding of the environment. That is, "our "dominion" over the universe should be understood more properly in the sense of responsible stewardship" (Pope Francis 2015).

Although the Church has often been characterized as rigid, it also has a duty to examine the "signs of the times" in the context of the Gospel. Such an approach speaks to Finnemore and Sikkink's characterization of a norm entrepreneur, which "must speak to aspects of belief systems or lifeworlds that transcend a specific cultural or political context" (Finnemore and Sikkink 1998, 907). As a norm entrepreneur, the Church facilitates the implementation and internalization of a norm cascade in other actors, thus transforming the global and local narratives on the environment and climate change. By re-conceptualizing the Church as a norm entrepreneur rather than just a civil society actor, scholars can more systematically incorporate the Church into studies of political development, change, and global agenda setting.

## Subsidiarity and Integral Ecology

The Church's public pronouncements regarding its understanding and approach to the international community and its willingness to work with state and non-state actors demonstrate its impact on the normative cycle. The Church's practice of subsidiarity and promotion of an integral ecology further embeds it in this process. As noted earlier, norm entrepreneurs focus on offering new cognitive frames to promote innovative ways of approaching and understanding an issue. What then is the new cognitive frame offered by the Catholic Church, and *how* does it propagate this new frame?

*The Compendium of the Social Doctrine of the Catholic Church* provides a systematic overview of the Church's fundamental social teaching. The doctrine clearly indicates its intent to provide a framework to "address appropriately the social issues of our day, which must be considered as a whole, since they are characterized by an ever greater interconnectedness, influencing one another mutually and becoming increasingly a matter of concern for the entire human family" (Pontifical Council for Justice and Peace 2004). The purpose of such action is guided by the "signs of the times" and subsidiarity. Together, they guide the Church's position on issues by considering changing social circumstances and recognizing the freedom and responsibility accorded to individuals and groups to handle such circumstances. Thus, to overcome the geographical and analytical divide between the Holy See (which represents the universal church) and the national Catholic Churches worldwide, the Catholic Church adopts the principle of subsidiarity to mediate between the national Churches and the Holy See as well as between the Holy See and the international community.

The Church's universality and uniformity is best understood as an entity present in areas throughout the world united by a single faith. Yet, the universal Church recognizes Catholic bishops to be national representatives who pastorally oversee their churches. Quoting from "Quadresimo Anno," the Compendium of the Social Doctrine of the Church recognizes the division of labor between the universal and national churches, stating:

> [I]t is an injustice and at the same time a grave evil and disturbance of right order to assign to a greater and higher association what lesser and subordinate organizations can do ... *On the basis of this principle, all societies of a superior order must adopt attitudes of help ("subsidium") — therefore of support, promotion, development — with respect to lower-order societies*. In this way, intermediate social entities can properly perform the functions that fall to them without being required to hand them over unjustly to other social entities of a higher level, by which they would end up being absorbed and substituted, in the end seeing themselves denied their dignity and essential place (italics in original).
> *(Pontifical Council for Justice and Peace 2004)*

Subsidiarity, therefore, recognizes that decision-making involves all of society and that the consequences of those decisions are interrelated.

Regarding the environment, Pope Francis advocated for a more holistic approach using the "principle of subsidiarity, which grants freedom to develop the capabilities present at every level of society, while also demanding a greater sense of responsibility for the common good from those who wield greater power" (Pope Francis 2015). Thus, as international conferences and organizations bring together political leaders, environmental experts, and specialized civil society groups, the Church's practice of subsidiarity asks individuals and families to be mindful of their environmental responsibility.

In addition to subsidiarity, the Church has consistently promoted the norm of integral ecology, and as a norm entrepreneur, it can help create not just a legal shift but, more importantly, a cultural shift. As noted by Pope Francis,

> We should not think that political efforts or the force of law will be sufficient to prevent actions which affect the environment because, when the culture itself is corrupt and objective truth and universally valid principles are no longer upheld, then laws can only be seen as arbitrary impositions or obstacles to be avoided.
> *(Pope Francis 2015)*

The politics of states, therefore, will not be enough to correct the negative effects of climate change. With a cultural, normative shift in understanding the environment, what is revealed is that

> The human environment and the natural environment deteriorate together; we cannot adequately combat environmental degradation unless we attend to causes related to human and social degradation... [W]e have to realize that a true ecological approach *always* becomes a social approach; it must integrate questions of justice in debates on the environment, so as to hear *both the cry of the earth and the cry of the poor*.
> *(Pope Francis 2015)*

From this perspective, the correlations are clear: a political culture of irresponsibility toward individual citizens is deeply tied to a culture of environmental neglect and exploitation.

## The Destruction of the Amazon: From Brazil's Colonial Past to Bolsonaro

How can an integral ecology change global society's approach to the Amazon? There is no denying the Amazon's significance to the health and future of the international community. What has become a point of tension, however, is *who* is responsible for the Amazon. Brazil's colonial past has been used to justify its resistance to opening itself to the international community. As early as the 1500s, European colonialists had ventured into the Amazon River Basin. By 1750, Spain and Portugal established the Treaty of Madrid, granting Portugal nearly all of what is modern-day Brazil's borders, including 60 percent of Amazonian territory (Casarões and Farias 2022, 57).

Over the years, state and international economic and political interests took turns dominating the Amazon. Ultimately, the "Amazon belonged to whomever had sovereign claim over it - only they were entitled to its riches, except for those non-nationals authorized by the government to reap some benefits" (Casarões and Farias 2022, 59). Although Brazil's colonization came to an end in the early 19th century, international concern and claim over the Amazon continued. In the late 1980s–early 1990s, then future US vice president, Senator Al Gore, claimed, "contrary to what Brazilians think, the Amazon is not their property, it belongs to all of us." Other European countries, such as France, argued that Brazil was incapable of preserving the Amazon and that "it was natural that third world countries surrendered their sovereignty over environmental protection issues" (Casarões and Farias 2022, 67).

Unsurprisingly, Brazil became increasingly suspicious of the international community's claims to protect the Amazon. The 1972 United Nations Conference on the Human Environment in Stockholm prioritized the Global North's concern with transnational pollution and the conservation of natural resources and biodiversity, which shaped a more self-interested Brazilian foreign policy on climate (Kiessling 2018, 388). In 1978, further attempts to keep the international community at bay included Brazil's support of a regional approach to protecting the Amazon (i.e. the 1978 Treaty of Amazon Cooperation). Here, the discussion on the Amazon shifted from "external intervention to national commitments of protecting the Amazon in line with environmental standards and global expectations" (Casarões and Farias 2022, 68).

As Brazil prioritized its national self-interest, it engaged in major infrastructure projects, such as deforestation, in order to further economically develop. Responding to international climate change agreements of the mid-late 1990s (i.e. the 1997 Kyoto Protocol), a Brazilian Ministry of Foreign Affairs official noted,

> Our question was: how much could we grant of our sovereignty, how much could we grant in our international decision making to supranational decision-making in development, energy and environmental policy? That was at the centre of our work with regard to climate change.
>
> *(quoted in Kiessling 2018, 398)*

The goal of Brazil's approach to international environmental agreements, therefore, was the preservation of national sovereignty and development.

Christopher Kiessling, a researcher on global governance of climate change, argued that Brazil's particular framing of the discussion on the environment is influenced by ideational and material factors. He noted that until 2000, Brazil's environmental policy was determined exclusively by government officials who "identified climate change as a problem of development and use of energy, NGOs clearly pointed to climate change as an environmental problem" (Kiessling 2018, 398). Thus, who is included (and excluded) in policymaking conversations appears to make a difference. As we will see later, the Church's voice in such conversations can help shape policy as well.

After 2000, it appears that the international focus on climate change had an impact on Brazilian foreign policy. It was during this time that Brazil created the Brazilian Forum of Climate Change and appointed Amazon activist Marina Silva as Minister of the Environment. Both helped to shift the domestic conversation to pay more attention to conservation and preservation and not just Brazil's development of the Amazon. After witnessing the international discussions on climate change, Fabio Feldmann, a federal deputy and first president of the Forum, encouraged then-Brazilian President Fernando Henrique Cardoso to be prepared for such talks. Feldmann recognized that while Brazil did not see climate change as a priority, the topic was already being discussed by various heads of state (Kiessling 2018, 400). Since discussing climate change was becoming a global norm, Brazil was compelled to follow suit.

Following the Forum's institutionalization and the inclusion of organized civil society, Brazil's traditional stance of disregarding the significance of regulating forest preservation on climate change and global climate governance was challenged (Kiessling 2018, 396, 400). Kiessling concludes that while Brazil was "inflexible in the non-inclusion" of forests in the climate issue prior to 2003, broadening the political space to involve alternative voices coupled with the Ministry of Environment's reframing of the climate issue helped to influence Brazilian foreign policy. In the 2005 meeting of COP 11, Marina Silva requested that the international community recognize the environmental importance of conserving tropical forests, which "enabled the possibility of receiving international financing for the implementation of national forestry programmes, changing a long stance in the subject by the Brazilian government" (Kiessling 2018, 403). Kiessling's work, therefore, highlights the significance of the constructivist approach to understanding Brazil's foreign policy on climate change. Discourse, ideas, and the reframing of issues can influence policymaking. Such an analysis only further supports the argument that the ideational influence of transnational, non-state actors such as the Catholic Church can play a role in shaping domestic and global norms on the environment. This role became more important following the 2019 Brazilian presidential election of far-right candidate Jair Bolsonaro, who initiated a rollback in the state's protection of the Amazon.

Despite these positive changes, which resulted in a historical ten-year decline in deforestation rates, Brazil's 2014 recession and the international demand for soy and beef led to an uptick of deforestation between 2013 and 2013 (Rossi 2014). As the world's top exporter of beef and soy (totaling more than $35 billion of both products in 2020), the high demand incentivized the renewed clearing of the Amazon. According to Yale's School of Forestry and Environmental Studies, cattle ranches account for nearly 80 percent of deforested Amazon land (Moriyama and Sandy 2019). During his 2019 presidential campaign, Bolsonaro empowered loggers, ranchers, and miners to transgress into the Amazon as he claimed Brazil's environmental policy "was suffocating the country," dismissed official data on deforestation, and vowed not "a square centimeter of land" would be designated for the Indigenous population (Moriyama and Sandy 2019).

The images of the Amazon burning drew international attention within the first year of Bolsonaro's presidential term. Until Bolsonaro took office in January 2019, no more than 3,000 square miles of Amazon land had been lost to deforestation ("Bishops Call Authorities in Brazil 'Arsonists' for Denuding Amazon Rainforests" 2021). That amount, however, has increased dramatically since his election. This increase in fire activity was perhaps mainly due to Bolsonaro's rollback of many of the prior efforts to curb deforestation in Brazil. Marina Silva, who, as environment minister in the mid-2000s, oversaw an 83 percent drop in deforestation between 2004 and 2012, stated, "Deforestation and fires have always been a problem, but this is the first time it has happened thanks to the discourse and activities of the federal government" (Moriyama and Sandy 2019). Bolsonaro referred to his policies not as a rollback but as a "reorientation" (i.e. dismantling) of Brazil's environmental protection policies, which had been progressively developing in the last forty years (de Araújo 2020, 2). For example, he proceeded to cut the environmental budgets[1] and attempted to dismantle the Ministry of Environment and Climate Change (MMA). The MMA is the ministry that oversees the Brazilian Institute of the Environment and Renewable Natural Resources (Ibama) and the Chico Mendes Institute for Biodiversity Conservation (ICMBio), which is solely responsible for the oversight of 75.9 million hectares designated for Federal Conservation.[2]

To preempt additional international backlash and possible infringement of Brazil's sovereignty, Bolsonaro mobilized the military to combat the fires and placed a 60-day ban on fires in the Amazon (Moriyama and Sandy 2019). However, based on the pretext of protecting Brazil's sovereignty,

Bolsonaro declined the Group of Seven's (G7) offer of millions of dollars of aid and proceeded to implement policies that would further weaken environmental protections (Roy 2022). Later, following additional criticism for threatening to undo the MMA, Bolsonaro enacted Decree No. 9471 in March 2019. With this, national agencies tasked with protecting the environment suffered major cuts, which affected their ability to monitor violations. The damage inflicted on the Amazon during Bolsonaro's presidency is palpable: over 12,800 square miles of the Brazilian Amazon were cleared – 60 percent more than in the three years prior (Nugent 2022).

**The Church's Response**

States and secular international organizations, however, are not the only players that can turn the political tide in Brazil; religious actors such as the Church have historically played a role in the region. According to a 2019 poll conducted by Pew Research Center, among the nations surveyed in Latin America, Brazil has a distinct relationship with religion, with 59 percent of those surveyed indicating they support religion having an important role in society. In 2021, Brazil's Institute for Climate and Society found that 43 percent of Brazilian voters assessed Bolsonaro's protection of the Amazon as "bad or very bad." These figures indicate that Brazilians would be receptive to the Church's advocacy for greater environmental responsibility. The Church, therefore, has the potential to make an impact domestically as much as it does internationally.

*From Aparecida to Laudato Si: Turning a Latin American Voice to a Global One*

While "Laudato Si" is often the point of focus when it comes to the Church's statement on the environment, the encyclical is heavily influenced by what transpired during the 2007 meeting of the Latin American bishops in Aparecida, Brazil, who gathered to discuss the future of the Church in the region. As a result of these meetings, they issued a Concluding Document, which addressed the significance of the Latin American region's contribution to the environment, problems of migration, trafficking, and the plight of the Indigenous. The bishops were guided by the concept of integral ecology and echoed Pope Benedict XVI's sentiment that "globalization must be led by ethics," especially given its current failure to meet the standards of human dignity. Using the method of "see-judge-act," the bishops urged others to "review critically one's life and work situation, reflect on these in the light of some Scripture passages, and with the support of others become empowered to take action to change and improve conditions" (Duncan 2020, 48). They noted that theology could interact with the social sciences; coupled with civil society and international organizations,

> the Church can engage in ongoing Christian interpretation and a pastoral approach to the reality of our continent, utilizing the rich legacy of the Church's social doctrine. It will thereby have concrete bases for demanding that those who are responsible for designing and enacting the policies that affect our peoples will do so in keeping with ethics, solidarity, and genuine humanism.
> 
> *(Conference of the Bishops of Latin American and the Caribbean 2007, 8.4 sec 403)*

Politics and policymaking, therefore, are a communal endeavor underscored by an ethical responsibility to the human family. Using an integral ecology approach, the Church serves as a constant reminder to go beyond mere utilitarian approaches to development.

Eight years after the Aparecida meetings, Pope Francis' groundbreaking encyclical, "Laudato Si," meaning "praise to you," was released on May 24, 2015. Building on the sentiments of the 2007 Aparecida, it was clear that "Laudato Si" would be a statement on the environment, as the title is borrowed from the "Canticle of Creatures," penned by St. Francis of Assisi, the patron saint of ecology. Released the same year the United Nations published its Sustainable Development Goals (SDGs) and the Paris Climate Change Conference was held, "Laudato Si" added a moral narrative to the mainly scientific discussions of the environment and climate change.

Using integral ecology as the basis for Laudato Si, Pope Francis was able to ground concern for the environment with the broader sociopolitical and economic issues that already engaged the Church. When tied to international relations, "Laudato Si" utilized the notion of "milieu goals," which is more expansive than "possessive goals" established by and for a particular state. According to international relations scholar Arnold Wolfers, "Milieu goals are of a different character. Nations pursuing them are out not to defend or increase possessions they hold to the exclusion of others, but aim instead at shaping conditions beyond their national boundaries" (quoted in Ferrara 2019, 2). Thus, "Laudato Si" tackled issues that affected integral ecology; such issues were not limited to the environment but also poverty and globalization.

"Laudato Si" reiterated that the Church does not take a narrow, snapshot view when considering how to view and handle an issue. Pope Francis states:

> When we speak of the "environment", what we really mean is a relationship existing between nature and the society which lives in it....We are

part of nature, included in it and thus in constant interaction with it. Recognizing the reasons why a given area is polluted requires a study of the workings of society, its economy, its behaviour patterns, and the ways it grasps reality...Strategies for a solution demand an integrated approach to combating poverty, restoring dignity to the excluded, and at the same time protecting nature.

*(Pope Francis 2015)*

Most importantly, Pope Francis argued that policy changes alone insufficiently remedy all the ills that plague society. Instead, a global norm of integral ecology can perhaps make a difference. "Laudato Si" pairs recommendations for structural change with a mental transition so that such structures can "question their own core commitments, and if necessary change themselves; to be something different, rather than just do different things" (quoted in Ferrara 2019, 16).

Pope Francis also pointed out that globalization has contributed to an international community that has become increasingly less communal, with near-sighted states lacking political will. All this hinders society from recognizing intergenerational solidarity, which must be considered "a basic question of justice, since the world we have received also belongs to those who follow us" (Pope Francis 2015). He promotes the Church's stance on an environmental education aimed at creating an "ecological citizenship," which leads to the cultivation of specific values that contribute to long-lasting rather than short-term political, social, and economic gains.

With the Aparecida, "Laudato Si," and the Synod for the Amazon, we see the Church's efforts, motives, and mechanisms of influence that Finnemore and Sikkink identified in norm entrepreneurs. In these three instances, the Church uses its local, regional, and global influence to change the norms in discussions regarding the environment by highlighting the moral narrative alongside the scientific language regarding climate change. In 2020, inspired by "Laudato Si," the Vatican's Dicastery for Promoting Integral Human Development launched the Laudato Si Action Platform. The Platform's goal is to "spark a cultural shift" and "sends a symbolic message that the global church is ready to do its part to take action on climate change" (Roewe 2021). Like the UN's SDGs, the Action Plan invited every member of the global community – not just states – to be more cognizant of an integral ecology. The platform provides participants plans and templates for action, which include steps towards "using renewable energy, reducing consumption of meat and single-use items; fostering ecological education and spirituality; advocating for sustainable development; and following ethical investment guidelines, including divestment from fossil fuels" (Roewe 2021). Unlike the SDGs, therefore, integral ecology pushes the envelope and does not just "propose targets to achieve" but aims higher for an ecological conversion.

Such a conversion is a "process, a way of becoming, and acting that integrates all dimensions of life, in order to address the global challenges of poverty, injustices and environmental degradation" (Deneulin 2021, 2288). There is a qualitative difference, therefore, between what the Church and the UN propose.

Using the lens of integral ecology, the Platform also provides tools and resources for individuals and communities (Dicastery for Promoting Integral Human Development 2021). Following its launch, over 4,000 church organizations and bodies signed up for the platform. According to Jose Aguto, Executive Director of Catholic Climate Covenant in the US, the Platform is "a tangible platform [for people] to feel like they are contributing to the solution" (Roewe 2021). The launch of the Laudato Si Platform demonstrates how the Church's efforts to shift the global norm on climate change can translate into action.

### *From "Laudato Si" to Synod for the Amazon*

As "Laudato Si" garnered international attention, the Latin American Church continued to carry on the spirit of environmental responsibility. Between 2017 and 2019, the Latin American Bishops convened the Synod for the Amazon, which was guided by the theme "New Paths for the Church and for an Integral Ecology." Continuing the work that began in Aparecida, the Synod was focused on how Latin America could put "Laudato Si" into action.

With the situation in the Amazon regarded as a "sign of the times," the three-week Synod in Rome gave the opportunity for Pope Francis to hear the voices from the Amazon. The Church felt the need to include the Amazon peoples as "principal dialogue partners" since they were the ones who were most affected by development projects. Central to the Synod was the Indigenous peoples' quest for "buen vivir" (good living), which can be considered an expression of an integral ecology. "Buen vivir" is understood as "a matter of living in harmony with oneself, with nature, with human beings and with the Supreme Being … here there are neither exclusions nor those who exclude, and here a full life for all can be projected" (The Synod of Bishops for the Pan-Amazon 2019). Thus, "buen vivir" is an integral approach that is coupled with "good acting," which involves the family, the community, and "responsible use of all the goods of creation" (The Synod of Bishops for the Pan-Amazon 2019). Once again, as a norm entrepreneur, the Church is proposing a different approach to how the Amazon is discussed and handled. They noted,

> Recognition and dialogue will be the best way to transform historical relationships marked by exclusion and discrimination. This local dialogue

in which the Church wants to be involved is at the service of life and of the 'future of the planet.'

*(The Synod of Bishops 2019)*

This is clearly a departure from the state-centered approach and international agreements on climate change. Remaining consistent with the Church's practice of subsidiarity, the Church created a space where Indigenous voices within the Amazon could be heard.

The Church's practice of subsidiarity was also evident in the convening and practice of the Synod. Church leaders were not the only ones involved in the Synod's Preparatory, Working, and Final Documents. Over the course of two years, such stakeholders included Indigenous people, theologians, and pastoral workers, totaling approximately 87,000 people from across the Amazon, along with 90 percent of the Amazon's bishops (Hansen 2019). These stakeholders were brought together by the Pan-Amazonian Church Network (REPAM), which was formed at the request of Pope Francis to support and protect Indigenous communities and their land (Duncan 2020, 50). REPAM conducted 300 listening sessions with participants from all nine countries in the Amazon, and half of the commission members who drafted the preparatory and working documents were from the area (Hansen 2019). With the Synod, the Church saw an "historic opportunity to distance itself from the new colonizing powers by listening to the Amazonian peoples and acting in a transparent and prophetic manner" (The Synod of Bishops for the Pan-Amazon 2019).

In the September 2019 Working Document, the Synod reiterated the Church's integral ecology by highlighting the interconnectedness of all life systems, the structures (economic, social, political, and cultural) that contribute to all forms of degradation, and the need for individual and collective self-reflexivity (Deneulin 2021, 2293). Assembly discussions identified nine challenges facing the Amazon: extractivist destruction, Indigenous people in voluntary isolation, migration, urbanization, family and community, corruption, integral health, integral education, and ecological conversion (The Synod of Bishops 2019). Outlined in Part II of the Working Document, the assemblies proposed detailed, multi-level plans that called upon governing bodies (local, national, and international), the Church, and local communities to work together for both changes in policies and culture.

The Synod also invited Carlos Nobre, a Brazilian earth scientist and 2007 Nobel Peace Prize winner, and 41 other international scientists and researchers. Together, they issued an 11-point plan titled "Scientific Framework to Save the Amazon" that echoed the sentiments of the Church's integral ecology. Included in this plan was the demand for the end of deforestation, a return of full funding for national enforcement and monitoring agencies (a point undoubtedly directed at the Bolsonaro administration), the pairing of

Indigenous knowledge of the Amazon with the work of local, regional, and international scientists, and a more transparent and rigorous approach to the monitoring of human rights abuses and land degradation in the Amazon (Scientists of the Amazon Countries and Global Partners 2019). These visionary efforts on the part of the Church sought to usher in a new approach to a perennial problem.

The Final Document issued in October 2019 reaffirmed the dangers facing the Amazon, the problematic nature of how economics and politics have been complicit in the "suffering and violence" inflicted on the Amazon, and the need for a broader and deeper dialogue with communities who have been ignored for too long. It concluded that to ensure the future of the Amazon, integral ecology "is the only possible path, because there is no other viable route for saving the region" (The Synod of Bishops for the Pan-Amazon 2019). According to Augusto Zampini, the Vatican's Director of Development and Faith, the Synod was committed to providing a concrete plan, which involved a development model for the Amazon communities and scientific institutions (Farand 2019). This indicates that the Church's influence is not limited to prayer or moral condemnations; it has both ideational and material resources that can help contribute to sociopolitical life.

The Synod, however, was criticized by both Bolsonaro and the Church's more conservative elements. Bolsonaro accused the Church of promoting a "leftist agenda," and in response to the Post Synod letter stated, "Pope Francis said yesterday the Amazon is his, the world's, everyone's…Well, the pope may be Argentine, but God is Brazilian," which was meant to reinforce Brazil's sovereignty over the Amazon (Staff and Agencies in Brasília 2020). While the state claimed to welcome the Synod and the Church's influence in Brazil, Brazil's intelligence agency (ABIN), Ministry of Foreign Affairs (ITAMARTY), and the Ministry of the Environment all were mobilized to "neutralize" the Church's sociopolitical impact and monitor the Synod's participants and meetings in the name of protecting Brazil's sovereignty (Phillips 2019). The political influence of the Synod was recognized as Ricardo de Costa, advisor to Bolsonaro's education agency, stated, "The clergy should worry about saving people's souls, not saving trees" (Gustin 2019). Although Bolsonaro is inconsistently dismissive of the Church's influence, his efforts against it raise the question of proportionality: the use of the state's security forces is clearly disproportionate, or Bolsonaro is up against something much more powerful than he would like to admit.

In advance of the Synod, the Church issued several letters defending its right and responsibility to showcase the issues of the Amazon and to present an alternative model to Bolsonaro's environmental plan. As a civil society actor, the Church is challenging the state's politics; as a norm entrepreneur, the Church called for a moral re-examination of development in the Amazon. In July 2020, nearly a third of Brazilian bishops (152 out of 450) signed a letter titled "Letter to People of God," which condemned Bolsonaro's handling of the "perfect storm" that has landed on Brazil: the COVID-19 pandemic, economic collapse, and political tensions within the state. The letter is a searing critique of Brazil's situation:

> The political choices that brought us here and the narrative that proposes complacency in the face of the excesses of the Federal Government, do not justify inertia and omission in combating the ills that befell the Brazilian people....Closing its eyes to the appeals of national and international entities, the Federal Government demonstrates omission, apathy and rejection by the poorest and most vulnerable in society, namely: the indigenous, quilombola, riverside communities, the populations of the urban outskirts, the tenements and the people who live in the streets, by the thousands, throughout Brazil.
> *(Sinodo dos Bispos do Brasil 2020)*

Following the release of the bishops' letter, nearly 1,600 Brazilian priests voiced their support for the bishops. Signatories to both the bishops' and priests' letters were criticized by other contingents within the Church for being too political, and with these letters, stymied dialogue with the state.

Over the course of its long institutional life, the Catholic Church's influence on politics has evolved from direct influence to recognizing that its moral essence and obligations limit its political role, particularly in the realm of policymaking. This limited political role has not hindered the Church in speaking in favor of moral values associated with human rights, social justice, and most recently, the environment. While those who signed the letters claimed to have not overstepped their bounds, the point of the letters remains the same: the raising of religious voices on behalf of the least fortunate in society, who appear to have been overlooked by the state. In October 2022, Bolsonaro lost his reelection bid to Lula da Silva, who had pledged to protect the Amazon during his presidential campaign. Between 2004 and 2012, it was Lula's Workers' Party that had cut the annual deforestation rate in the Amazon by 80 percent and, in his re-election victory speech, announced, "Now, we are going to fight for zero deforestation" (Nugent 2022). It would appear then that Brazil may soon return to a path that involves greater environmental responsibility.

### Conclusion: The Moral and Not Merely Political Obligations of the Catholic Church

In a 2022 survey commissioned by Brazil's Institute for Climate and Society, 62 percent of the 3,000 Brazilians surveyed (i.e., 3 out of 4) responded that they would vote for a presidential candidate who had a specific plan to protect the Amazon. Furthermore, the "argument that economic development is only viable with deforestation has also collapsed: 70% of the voters believe that protecting the forest is important for development" (Lobo 2022). Brazilian businesses are also concerned with the impact of global environmental concerns on their image and, therefore, earning potential. According to Blairo Maggi, a billionaire soybean producer and former agriculture minister, "Did we have our image harmed? Yes. Can we recover it? Yes. The government has to align its discourse to what the world wants" (Moriyama and Sandy 2019). It is exactly this last point – "what the world wants" – that the Church's role as a norm entrepreneur is evident.

The case of the Amazon demonstrates the tension between the global commons and national sovereignty. Who has legitimate control and the ultimate say over a territorial area that plays a crucial role in the physical, social, and economic health of the international community? In his opening message to the March 2019 international conference titled "Religions and the Sustainable Development Goals," Cardinal Turkson outlined the role of the Catholic Church and of religion, more broadly, in international affairs. The conference's intention was to issue moral exhortations but also create concrete solutions catalyzed by a moral imperative. Mirroring the structure of the 2030 UN's Agenda for Sustainable Development, focusing on the five P's of People, Planet, Prosperity, Peace, and Partnership, the goal of the conference was to show how religion's key components should and can contribute to political life by encouraging an attitudinal and existential shift of one's views toward solidarity and responsibility.

In mirroring the structure, the conference was meant to highlight the ethical and moral dimensions that are often preempted by technocratic-focused solutions that are generally devoid of religion. In the case of the Amazon, the Church utilizes its resources to shift the conversation from one of *control over the Amazon* to one of *responsibility to the Amazon*. According to Bishop Joaquín Humberto Pinzón of the Colombian Amazon, the attacks on the Synod for simply raising awareness about the Amazon indicate the dissatisfaction of political and economic actors: "It doesn't suit them – neither the politician, nor the business people, not the owners of the big mining companies" (Phillips 2019). As a civil society actor and norm entrepreneur, the Church can broaden the local, domestic, and global discussion beyond the interests of political and economic elites. Perhaps it was not coincidental, therefore, that the post-Synodal letter was issued around the same time the Vatican welcomed to Rome Bolsonaro's political opponent and eventual 2022 presidential successor, former Brazilian president Luiz Inácio Lula da Silva. Lula received a papal blessing, and the two discussed "prospects for a more just and fraternal world" (Staff and Agencies in Brasília 2020).

Once dominated by Catholicism, Brazil is now home to the world's second-biggest Protestant population, who mainly follow the Universal Church of the Kingdom of God, a neo-charismatic evangelical church (Llywelyn 2022). Evangelical Protestants make up 22 percent of Brazil's population and are the fastest-growing religious demographic, whereas Catholics have experienced a decline down to 53 percent in 2019 from 60 percent in 2010 (Polimédio 2019). This religious shift will undoubtedly affect the Church's presence in Brazil as it struggles to compete for adherents. In its efforts to maintain its numbers, the Synod for the Amazon also discussed ways in which the Church could increase its presence in the region. Some of the topics included the ordination of married priests, allowing more prominent roles for women, and the incorporation of Indigenous concepts into Church theology (Gustin 2019). Such topics have been considered controversial by more conservative factions within the Church. The outcome of those discussions may overshadow the Church's broader message and efforts for ecological conversion in the region. The pressing issue of Catholicism in decline may also redirect the Church's attention to not prioritizing the environment in its discourses.

If the Church is to remain true to the concept of the integral ecology, however, the environment does not ever really recede from its focus, even though it may appear so to others. Instead, it is through integral ecology that the Church carries out its mission, and, therefore, no one issue is ever left to fall by the wayside. Furthermore, by contextualizing the Church as a norm entrepreneur, scholars and practitioners will have a better understanding of the Church's role in the practice and study of international relations.

> When it comes to moral agency, the normative approach to international relations insists on duties and obligations ("something must be done"), whereas international history elaborates on blame and accountability ("never again"). However, in both cases, failure to indicate the moral agents involved makes such refrains meaningless.
> *(Ferrara 2019, 12)*

Contextualizing the Church as a norm entrepreneur that promotes integral ecology helps to fill this theoretical and practical gap.

## Notes

1 For example, Ibama's 2020 budget for inspections was 25.3 percent less than in 2019, and the Federal Conservations Units' 2020 budget was 32.7 percent less than in 2019.

2 (de Araújo 2020, 3). MMA = Ministério do Meio Ambiente e Mudança do Clima; Ibama = Instituto Brasileiro do Meio Ambiente e dos Recursos Naturais Renováveis; ICMBio = Instituto Chico Mendes de Conservação da Biodiversidade.

## Bibliography

Araújo, Suely Mara Vaz Guimarães de. 2020. "Environmental Policy in the Bolsonaro Government: The Response of Environmentalists in the Legislative Arena." *Brazilian Political Science Review.* 14(2): 1–20. https://doi.org/10.1590/1981-3821202000020005.

"Bishops Call Authorities in Brazil 'Arsonists' for Denuding Amazon Rainforests." 2021. La Croix International. December 13, 2021. https://international.la-croix.com/news/environment/bishops-call-authorities-in-brazil-arsonists-for-denuding-amazon-rainforests/15358.

Bolsonaro, Jair. 2019. "Speech by Brazil's President Jair Bolsonaro at the Opening of the 74th United Nations General Assembly." September 24, 2019. https://www.gov.br/mre/en/content-centers/speeches-articles-and-interviews/president-of-the-federative-republic-of-brazil/speeches/speech-by-brazil-s-president-jair-bolsonaro-at-the-opening-of-the-74th-united-nations-general-assembly-new-york-september-24-2019-photo-alan-santos-pr.

Casarões, Guilherme, and Déborah Barros Leal Farias. 2022. "Amazon and the International Order: From Promise to Peril." *Journal of International Affairs.* 75(1): 55–21.

Conference of the Bishops of Latin American and the Caribbean. 2007. "Concluding Document: V General Conference of the Bishops of Latin America and the Caribbean. Aparecida, May 13–31, 2007." https://www.celam.org/aparecida/Ingles.pdf.

Deneulin, Séverine. 2021. "Religion and Development: Integral Ecology and the Catholic Church Amazon Synod." *Third World Quarterly.* 42(10): 2282–2299. https://doi.org/10.1080/01436597.2021.1948324.

Dicastery for Promoting Integral Human Development. 2021. "Laudato Si' Action Platform." Laudato Si' Action Platform. 2021. https://laudatosiactionplatform.org/about/.

Duncan, Bruce. 2020. "The Amazon Synod: Putting Laudato Si' into Action." *Interface Theology.* 6 (December): 47–65.

Farand, Chloé. 2019. "Catholic Church Denounces 'Attacks' on Amazon People and Forest." Climate Home News. October 29, 2019. https://www.climatechangenews.com/2019/10/29/catholic-church-denounces-attacks-amazon-people-forest/.

Ferrara, Pasquale. 2019. "Sustainable International Relations. Pope Francis' Encyclical Laudato Si' and the Planetary Implications of 'Integral Ecology.'" *Religions.* 10(8): 466. https://doi.org/10.3390/rel10080466.

Finnemore, Martha, and Kathryn Sikkink. 1998. "International Norm Dynamics and Political Change." *International Organization.* 52(4): 887–917. https://doi.org/10.1162/002081898550789.

Gustin, Georgina. 2019. "As Amazon Fires Burn, Pope Convenes Meeting on the Rainforests and Moral Obligation to Protect Them." *Inside Climate News* (blog). October 6, 2019. https://insideclimatenews.org/news/06102019/pope-amazon-forest-fires-synod-bolsonaro-religion-catholic-church-climate-change/.

Hansen, Luke. 2019. "In the Amazon, Pope Francis is Setting the Agenda for a New Kind of Synod." America Magazine. September 12, 2019. https://www.americamagazine.org/faith/2019/09/12/amazon-pope-francis-setting-agenda-new-kind-synod.

Kiessling, Christopher Kurt. 2018. "Brazil, Foreign Policy and Climate Change (1992-2005)." *Contexto Internacional*. 40(2): 387–408. https://doi.org/10.1590/s0102-8529.2018400200004.
Llywelyn, D. 2022, April 3. Global Christianity: The Future of the Catholic Church. *Institute for Advanced Catholic Studies at USC*. https://dornsife.usc.edu/iacs/2022/04/30/global-christianity/
Lobo, Felipe. 2022. "The Defense of the Amazon Is the Reason to Vote for President for 62% of Brazilians." iCS. September 20, 2022. https://climaesociedade.org/en/the-defense-of-the-amazon-is-the-reason-to-vote-for-president-for-62-of-brazilians/
Moriyama, Victor, and Matt Sandy. 2019. "'The Amazon Is Completely Lawless': The Rainforest After Bolsonaro's First Year." *The New York Times*, December 5, 2019, sec. World. https://www.nytimes.com/2019/12/05/world/americas/amazon-fires-bolsonaro-photos.html.
Nugent, Ciara. 2022. "With Bolsonaro Gone, Lula Could Save Amazon Rainforest | Time." October 31, 2022. https://time.com/6226932/lula-win-amazon-climate-change-brazil/.
Phillips, Tom. 2019. "Bolsonaro Targets the Catholic Church over its 'leftist Agenda' on the Amazon; Gathering at the Vatican Has Triggered a Political Storm in Brazil as Bishop Denies Undermining the Government." *The Guardian (London)*, September 23, 2019. https://www.theguardian.com/world/2019/sep/23/bolsonaro-targets-the-catholic-church-over-its-leftist-agenda-on-the-amazon.
Polimédio, Chayenne. 2019. "How Evangelical Conservatives Are Gaining Power in Brazil." *Foreign Affairs*, March 7, 2019. https://www.foreignaffairs.com/articles/brazil/2019-03-07/how-evangelical-conservatives-are-gaining-power-brazil.
Pontifical Council for Justice and Peace. 2004. "Compendium of the Social Doctrine of the Church." April 2, 2004. http://www.vatican.va/roman_curia/pontifical_councils/justpeace/documents/rc_pc_justpeace_doc_20060526_compendio-dott-soc_en.html.
Pope Francis. 2015. "Encyclical Letter Laudato Si' of the Holy Father Francis on Care for Our Common Home." http://w2.vatican.va/content/dam/francesco/pdf/encyclicals/documents/papa-francesco_20150524_enciclica-laudato-si_en.pdf.
Roewe, Brian. 2021. "Official Launch of Laudato Si' Action Platform Offers Catholics Concrete Steps toward Sustainable Lifestyles." National Catholic Reporter. November 14, 2021. https://www.ncronline.org/earthbeat/justice/official-launch-laudato-si-action-platform-offers-catholics-concrete-steps-toward.
Rossi, Marina. 2014. "Deforestation in Brazil Increases for the First Time in a Decade." EL PAÍS. November 13, 2014. https://english.elpais.com/elpais/2014/11/13/inenglish/1415894133_192550.html.
Roy, Diana. 2022. "Deforestation of Brazil's Amazon Has Reached a Record High. What's Being Done?" Council on Foreign Relations. August 24, 2022. https://www.cfr.org/in-brief/deforestation-brazils-amazon-has-reached-record-high-whats-being-done.
Scientists of the Amazon Countries and Global Partners. 2019. "Scientific Framework to Save the Amazon." September 30, 2019. http://secretariat.synod.va/content/sinodoamazonico/en/news/scientific-framework-to-save-the-amazon-by-scientists-of-the-ama.html.
Sínodo dos Bispos do Brasil. 2020. "Carta Ao Povo de Deus." Conselho Pastoral dos Pescadores. http://cppnacional.org.br/sites/default/files/Carta%20ao%20Povo%20de%20Deus%20VF%2011.08.2020%20com%201520.pdf.
Snyder, Jack, Ed. 2011. *Religion and International Relations Theory*. New York: Columbia University Press.
Staff and Agencies in Brasília. 2020. "Bolsonaro Attacks Pope Francis over Pontiff's Plea to Protect the Amazon." *The Guardian*, February 13, 2020, sec. World news.

https://www.theguardian.com/world/2020/feb/13/brazil-jair-bolsonaro-pope-francis-amazon.

The Synod of Bishops. 2019. "Instrumentum Laboris: The Working Document for the Synod of Bishops." September 1, 2019. http://secretariat.synod.va/content/sinodoamazonico/en/documents/pan-amazon-synod–the-working-document-for-the-synod-of-bishops.html.

The Synod of Bishops for the Pan-Amazon. 2019. "Final Document of the Amazon Synod." October 26, 2019. http://secretariat.synod.va/content/sinodoamazonico/en/documents/final-document-of-the-amazon-synod.html.

United States Conference of Catholic Social Bishops. 2005. "Seven Themes of Catholic Social Teaching." https://www.usccb.org/beliefs-and-teachings/what-we-believe/catholic-social-teaching/seven-themes-of-catholic-social-teaching

# 7
## NOW WHAT?
### Implications for Academics, Policymakers, and Practitioners Across and Between Religious and Secular Contexts

*Kalzang Dorjee Bhutia, Amy Holmes-Tagchungdarpa, Lan T. Chu, and Youssef Chouhoud*

In our case studies across different religious traditions around the world, we have explored how different religious practitioners both individually and institutionally respond to the challenges brought by climate change. Climate change is understood in a variety of ways, but there is no singular Indigenous, Muslim, Christian, or Buddhist approach, as each of these traditions encapsulate tremendously diverse perspectives and worldviews. These different communities also understand the environment and their relationship to it in very different ways. Despite these very diverse perspectives, our case studies have demonstrated how localized approaches to climate change can lead to positive change and particularly the empowerment of local communities with a sense of agency in the face of rapid change. We have argued that religious perspectives on both individual and institutional levels demonstrate the importance of religion for developing response to climate change and its many impacts. Rather than emphasizing a binary between religion and science, where science is emphasized as the only solution to climate change, personal and institutional forms of religion provide people with alternative ways to understand responsibility and care for the environment. Here, we outline implications that can be found across our case studies and recommendations for ways that religious practice and discourse can be acknowledged in responding to climate change.

### Implication 1: Creating Conversations across Religious and Scientific Communities and Discussions

At present, scientists often resist input from religious communities and assume a binary between religious and secular action and understanding; religious communities may also not seek out input from scientists. This prevents

DOI: 10.4324/b23076-7

the incorporation of powerful religious worldviews that inspire different communities to make decisions that have positive environmental outcomes. A recognition of both scientific and religious viewpoints as sources of truth for different people can lead to discussions and collaborations based on mutual respect and increase understanding and the potential for change.

An example of an initiative that has bridged scientific recommendations about the significance of trees to avoid the negative impacts of deforestation was discussed in the case of Thailand, where environmental monks ordain trees as a way to promote environmental care (Darlington 2013). In the region of West Sikkim where Kalzang Dorjee Bhutia is from, local community members – spanning ethnic and religious identities – come together frequently for tree-planting initiatives. The Sikkim State Government has attempted to co-opt these initiatives, framing them as being part of "green" state initiatives, and the Forest Department also supports this with input from local forest scientists. However, tree planting drives began much earlier and have taken place for many years; Bhutia's older family members and friends recall doing this in the 1920s and 30s. There are many different religious motivations for undertaking tree plantations, including environmental care and concern for deities resident in local environments that are recognized by Indigenous, Buddhist, and Hindu cosmologies in the region (Acharya and Ormsby 2017; Bhutia 2021). Sonam Wangchuck Bhutia, a social worker and lama at Pemayangtse Monastery, has a yard full of rhododendrons and orchids that he plants every year. Occasionally, the State provides money to pay for the seedlings, but mostly he donates them himself. He does not frame his commitment to planting trees in terms of state desires or beautification, but instead as a responsibility to maintain his relationships and care for the local protector deities. He recognizes that these deities may have different names or be thought about differently by his neighbors in nearby villages that practice different religions, but he still sees it as the shared responsibility of the community to care for them and to collaborate to do so. These tree-focused initiatives both demonstrate how scientists and religious practitioners can collaborate, even when their understanding of environmental care may be different.

### Implication 2: Less Simplistic Understandings of Religion Are Needed

Religion operates differently across different elements of culture and society and influences how people relate to their environment. Religion itself is a complicated category and is understood differently in different local and global settings. In Chapter 1, we have seen how Indigenous communities may not use the term religion but instead carry out practices that incorporate environmental care and emphasize relatedness and are motivated by cosmologies and worldviews that may be part of several religions or other frameworks entirely. In Chapter 2 on Islam, we saw how an apparently single tradition can contain a variety of manifestations and compel people to act based on legal and ethical frameworks that may be pervasive beyond what may be assumed to be religious boundaries.

Religion can also be very powerful as a way to motivate action. Religions should not just be understood in their institutional forms and histories, which may be tied to forms of colonialism and oppression. Expanding the definition of religion means taking into account people's worldviews, ethical standpoints, value systems, and understandings of how they relate to their communities and the broader world. Some people may not identify as any religion or as members of multiple religious communities. Understanding religion as complicated and not fixed to specific definitions allows for different opportunities for action and engagement to emerge outside of limited frameworks.

## Implication 3: Personal Belief and Action Is as Important as Institutional and State Policy

The 2015 UN Climate Change Conference (COP21) had a huge impact on climate change policy, culminating in the signing of the international treaty on climate change, which is known as The Paris Agreement (United Nations 2023). In the months leading up to and after the Conference, many religious organizations renewed or released new statements relating to their respective positions on religious responsibilities towards the environment (Yale Forum on Religion and Ecology 2023a). These statements include Pope Francis' now-famous 2015 encyclical *Laudato si'* (Pope Francis 2015). The appearance of these statements has led to a new period of interfaith dialog and discussions around climate change. However, since most religious organizations and communities are not under the purview of a global authority (the Vatican is a clear exception), the ability of these statements and meetings to transform policy and action in different places and contexts has been debated (Sadouni 2022; Krantz 2023).

It is important to see how religious statements from institutions only outline some perspectives and viewpoints. Academics and policymakers should strive to understand and uplift the power of religious actors to act as individuals instead of assuming they only act according to institutional authority. Just as religious actors do not always vote along lines set by their institutional authority, they can also make a variety of decisions – whether to recycle, embrace green energy, or use electric cars – without consideration of what their institutions tell them to do. Religious actors are complicated, and many other factors contribute to the making of an individual's identity. Religious institutions and the State are not the only actors who can bring about awareness of, and action in behavior, climate change mitigation. Individuals may not always agree with their religious institutions or with state policy, but their decision making in terms of consumption and different forms of action can have a meaningful impact on carbon emissions and other impacts of climate change.

### Implication 4: Including Local Worldviews and Concerns into Decision Making in Meaningful Ways

Beyond individuals, local communities may practice their religions in different ways and have diverse understandings of their relationship with the environment in both urban and rural settings. These understandings have been formed over time, based on specific ecological and social contexts. Effective policy and actions need to take into account these differences and avoid one-size-fits-all strategies and forms of domination and imperialism that prevent meaningful change by reifying historical power imbalances.

Even within specific traditions, we can see the influence of different environments and how that has impacted how communities respond to environmental challenges. For example, Chapter 3 about Buddhism demonstrated how diverse Buddhist perspectives on care for animals can be. While the sponsorship of freeing animals intended for slaughter to obtain merit is popular across different parts of the Buddhist world (Shiu and Stokes 2008), it is carried out in very different ways between Coastal China and Vietnam and the highlands of the Tibetan plateau. These different ecological settings have directly contributed to debates about the potential for unforeseen negative consequences of these practices and the way they may actually lead to loss of biodiversity, as opposed to the preservation of species. Conversations with local communities are vital for discussions about policy and practice by both State and religious actors that can create favorable alternatives (such as fundraising for nonhuman animal neutering) or revive traditional forms of sustainability (such as freeing cows to be cared for by local communities in appropriate rural settings).

### Implication 5: Understand How the Environment Can Be Understood in Complicated Ways across Religious and Cultural Perspectives

There is no one definition of the environment, and religious actors and communities can understand the environment in different ways. These ways do not always map neatly onto the way conservationists and scientists talk about religion. Religious actors and communities may already be engaged in acts of environmental care that they talk about in different ways – as acts of service, for example.

With regard to Catholicism, the Catholic Church's use of integral ecology is an effort to motivate others to view the environment more than simply as a natural resource that needs to be saved from exploitation. Instead, integral ecology is linked to human development and serves as a "new paradigm of justice that connects the exercise of care for nature with the exercise of justice for the most impoverished and disadvantaged on earth" (Desierto and Schnyder Von Wartensee 2021, 1529). With regard to the Amazon, therefore, it is not solely a material resource but a broadly human one as well. Going beyond self-interested states and international human rights paradigms that focus on the recognition of rights and states' duties, the Church "places the reparative and redressive role of Integral Ecology at the forefront of the dialogues on environment, ecological stewardship, national and local actions, and the exercise of self-determination by the peoples of the Amazon region" (Desierto and Schnyder Von Wartensee 2021, 1531). As noted in Chapter 6, the Church acts as a norm entrepreneur, inspiring states and non-state actors to recognize and adhere to the norm of global integral ecology.

### Implication 6: Listen to Indigenous Communities and Avoid Stereotypes

Indigenous communities have had longstanding relationships with the environments they live in but are all too often not included in the development of climate policy. The idea of Indigenous people as inherently in tune with the environment is a simplistic stereotype that ignores how different communities have adapted to different challenges over time. Nuanced and thoughtful discussion between Indigenous, religious, and political actors needs to be sustained to avoid tokenism (Chakraborty and Sherpa 2021).

An example of a well-intended yet problematic incorporation of Indigenous cosmology into climate policy took place in 2009 in Bolivia, where the government introduced a new Constitution that aimed to protect nature, which was personified as *Pachamama*, Mother Earth. As legal scholars Paola Villavicencio Calzadilla and Louis J. Kotzé noted, the Constitution does not "recognize nature as the bearer or rights," but instead argues for "the right to a healthy, protected, and balanced environment for all people" (Villavicencio Calzadilla and Kotzé 2018, 401). Since the passing of the Constitution, a number of laws have been developed that are intended to support the Constitution (Villavicencio Calzadilla and Kotzé 2018, 404). However, the effectiveness of enshrining these ideas in the Constitution and law has been questioned due to tensions with the Constitution around the use of natural resources. This has led scholars to argue that the protection of *Panchamama* may be more "on paper" than actual, and more widespread changes are needed to actualize the promise of environmental justice on national levels (Villavicencio Calzadilla and Kotzé 2018, 423–424).

## Implication 7: Climate Change Has Many Impacts – Take into Account These Different Impacts in Developing Action and Policy Beyond States

State-led initiatives against climate change are hard to enact and complicated by the fact that climate change is not a localized, national-level phenomenon. While it is important to avoid climate reductionism, climate change does have significant interconnected impacts in different social, political, and ecological contexts. Climate change leads different communities to seek out new economic opportunities, which may also pose new forms of environmental degradation; climate change contributes to migration and the need for support and new infrastructure for migrants. Policymakers, scientists, and religious practitioners need to think about these different impacts and effects and consider holistically different responses that emphasize collaboration between different states, institutions, and individuals.

An example of this has been the mobilization around eco-anxiety, which is also called climate anxiety. The emergence of awareness of climate anxiety is traced to the first decades of the twenty-first century. In 2017, the American Psychological Association (APA) released a report on "Mental Health and Our Changing Climate: Impacts, Implications, and Guidance" (APA 2017). In the report, researchers discussed how the physical toll of climate change is well established but mental health is less recognized. They argued,

> It is time to expand information and action on climate and health, including mental health. The health, economic, political, and environmental implications of climate change affect all of us. The tolls on our mental health are far reaching. They induce stress, depression, and anxiety; strain social and community relationships; and have been linked to increases in aggression, violence, and crime….
>
> To compound the issue, the psychological responses to climate change, such as conflict avoidance, fatalism, fear, helplessness, and resignation are growing.
>
> *(APA 2017, 4)*

Around this time period, more initiatives to help people respond to eco-anxiety began to appear. Some of the most well known are those associations with religious, and particularly interfaith, cross-denominational, or spiritually open teachers, groups, and initiatives.

One prominent example is the Loka Initiative at the University of Wisconsin, Madison. The Loka Initiative defines its mission to be "to support faith-led environmental and climate efforts locally and around the world by helping build capacity of with leaders and cultural keepers of Indigenous traditions, and by creating new opportunities for projects, partnerships, and public outreach" (Loka Initiative 2023). The Loka Initiative is headed by Dekila Chungyalpa, a climate scientist and educator originally from the eastern Himalayan Indian state of Sikkim. In August 2023, the Loka Initiative held the Resilience in The Anthropocene (RITA) online summit (RITA 2023). Over three days, the summit featured a variety of speakers from different faith and cultural communities speaking about climate change in their own communities, the adverse impacts of climate change on marginalized communities, and sharing resources for promoting wellbeing in response to the many changes brought by climate change. Importantly, RITA was not framed as a Buddhist event, even though it did feature a number of Tibetan and American Buddhist teachers. Instead, it included a variety of approaches from psychology, environmental justice, and ethnic studies (RITA 2023).

RITA is just one example of many similar events and resources that are appearing in a variety of different settings: online, on university campuses, and through local community organizations around the world. The Yale Forum on Religion and Ecology has compiled a list of resources (Yale Forum on Religion and Ecology 2023b), demonstrating how widespread discussions of eco-anxiety are and how they are frequently linked to religion and spirituality as resources for response. These resources demonstrate the potential for interconnected efforts to deal with the cumulative, connected impacts of climate change that can operate effectively across different contexts.

**Bibliography**

Acharya, Amitangshu, and Alison Ormsby. 2017. "The Cultural Politics of Sacred Groves: A Case Study of *Devithans* in Sikkim, India." *Conservation & Society*. 15(2): 232–242.

American Psychological Association (APA). 2017. https://www.apa.org/news/press/releases/2017/03/mental-health-climate.pdf (Accessed October 15, 2023)

Bhutia, Kalzang Dorjee. 2021. "Trees as Village Protectors, Guru Rinpoche's Wayfinders and Adopted Family Members: Arboreal Imagination, Agency and Relationality in Sikkim." *Worldviews: Global Religions, Culture, and Ecology*. 25(2): 151–170.

Chakraborty, Ritodhi, and Pasang Yangyee Sherpa. 2021. "From Climate Adaptation to Climate Justice: Critical Reflections on the IPCC and Himalayan Climate Knowledges." *Climatic Change*. 167. doi.org/10.1007/s10584-021-03158-1

Darlington, Susan. 2013. *Ordination of a Tree*. Albany: State University of New York Press.

Desierto, Diane A., and Ilaria Schnyder Von Wartensee. 2021. "The Right to Development, Integral Human Development, and Integral Ecology in the Amazon." *The International Journal of Human Rights*. 25(9): 1525–1542.

Krantz, David. 2023. "Climate and Covenant: A Case Study of the Functions, Goals, and Tensions of Faith at the 23rd Conference of Parties to the United Nations Framework Convention on Climate Change." In *Religious Environmental Activism*, edited by Jens Koehrsen, Julia Blanc and Fabian Huber. London: Routledge, 282–302.

Loka Initiative. 2023. Loka Initiative (front page). https://centerhealthyminds.org/programs/loka-initiative (Accessed November 1, 2023)

Pope Francis. 2015. *Laudato si'*. https://www.vatican.va/content/francesco/en/encyclicals/documents/papa-francesco_20150524_enciclica-laudato-si.html (Accessed August 1, 2023)

RITA. 2023. https://www.ritasummit.org (Accessed November 1, 2023)

Sadouni, Samadia. 2022. *Religious Transnationalism and Climate Change*. Cham: Palgrave Macmillan.

Shiu, Henry, and Leah Stokes. 2008. "Buddhist Animal Release Practices: Historic, Environmental, Public Health and Economic Concerns." *Contemporary Buddhism*. 9(2): 181–196.

United Nations. 2023. *The Paris Agreement*. https://unfccc.int/process-and-meetings/the-paris-agreement (Accessed November 1, 2023)

Villavicencio Calzadilla, Paola, and Louis J. Kotzé. 2018. "Living in Harmony With Nature? A Critical Appraisal of the Rights of Mother Earth in Bolivia." *Transnational Environmental Law*. 7(3): 397–424.

Yale Forum on Religion and Ecology. 2023a. *Climate Change Statements*. https://fore.yale.edu/Climate-Emergency/Climate-Change-Statements-from-World-Religions (Accessed November 1, 2023)

Yale Forum on Religion and Ecology. 2023b. *Eco-Anxiety Resources*. https://fore.yale.edu/Resources/Eco-anxiety-Resources (Accessed November 1, 2023)

# INDEX

Note: Page numbers in *italics* refer to figures.

Ackerman, Thomas 8
Affected Citizens of the Teesta (ACT) 19
*African Perspectives on Religion and Climate Change* (Chitando) 6
*ahiṃsā* (nonviolence) 49, 54
Al-Musnad 29
*amana* 27
Amazon 8–9, 61, 85; deforestation 61–62; destruction of 65–75
American Psychological Association (APA) 86
anthropocentrism 29; of Abrahamic religions 17
anti-dam activism 18
Appleby, Scott 5
ArabBarometer 36, *38*, 41

Barstow, Geoffrey 55
Bhutia, Kalzang Dorjee 14, 46, 57–58, 82
Bhutia, Sonam Wangchuck 82
Blue Sky and Green Earth of One Mind 48
Bodhi, Ven Bhikkhu 44
Bodhisattva 48, 59
Bolsonaro, Jair 8–9, 61–62, 65, 68–69, 75
Brazil 61; colonization 66; environmental protection policies 68–69; intelligence agency (ABIN) 74; Ministry of Foreign Affairs (ITAMARTY) 74; Ministry of the Environment 74
Brazilian foreign policy 67
Brazilian Forum of Climate Change 67
Brown, Marie Alohalani 16, 20
Brown, Nathan 30
Buddhism 4, 8, 13, 44, 84; Buddhist practitioners 4; influenced vegetarianism 55; rituals 18
Buddhist understandings of human-environmental relationships 44–47; climate change and 47–48; countercultural religion 45; environment in practice 49–59; forest in northern Thai Buddhism 52–54; four noble truths 45; life release as human-nonhuman animal ethics 49–52; movements against plastic waste in the Himalayas 56–57; ordaining trees 52–54; rituals to counter the melting glaciers 57–58; traditional knowledge in 57–58; vegetarian movements in Taiwanese Buddhist communities 55–56

Calzadilla, Paola Villavicencio 85
Cardoso, Fernando Henrique 67
Carlson, Ervin 22
Catholic Church 8–9, 62, 84; environmental responsibility 65; as

norm entrepreneur 62–63; social teaching 64; subsidiarity and integral ecology 64–65
Catholicism 84
Catholic Social Teaching 63
Chakraborty, Ritodhi 12
Chasek, Pamela 6
Cheng Yen 55
Chettri, Mona 12, 19
Chitando, Ezra 6
Christianity 4, 8, 13, 17
Christian practitioners 4
Chronbach's alpha 40
Chungthang, Sikkim 12
civil society 26
Climate Action Tracker project 34
climate anxiety 86
climate change 2–4, 81, 86; and dams 13; Himalayas and 12; and Indigenous communities 4, 13–14; intra-religious responses to 7; probability of reporting 39; traditional knowledge in 57–58
*Climate Change, Religion, and Our Bodily Future* (LaVasseur) 6
*Climate Politics and the Power of Religion* (Berry) 6
climate reductionism 2
cock-eyed optimists 8
$CO_2$ emissions 31
*Compendium of the Social Doctrine of the Catholic Church, The* 64
constructivism 9, 62
COVID-19 pandemic 75
creation care proponents 8
critical political ecology 6

Dalai Lama 46
damming the rivers of Sikkim 13
*Dar-el-ifta'* 31
Darlington, Susan 52–54
Day of Judgment 29
deforestation 61, 68, 82; policies (Bolsonaro) 9
denialists 8
*Dharma Gaia* 47
*Dharma Rain* 47
Dien, Mawil Izzi 27
Downie, David 6
Dzongu 18

Eckersley, Robyn 6
*Ecodharma* 48
ecological citizenship 71

ecological crisis 4
*ecosattva* path 48
Egypt 31, 40
Elverskog, Johan 45, 54
end-time militants 8
Engaged Buddhism 47
environment 2
environmentalism 14
environmental monks 53
extractivist destruction 73
extravagance *(tabdhir)* 27

*fatwa* (religious ruling) 31
Feldmann, Fabio 67
Ferrara, Pasqual 4
Finnemore, Martha 62–63
food sovereignty 22
Frazer, Owen 5
Frazer and Friedli's model 5
Friedli, Richard 5
Fujikane, Candace 20

Gade, Anna 6
Gagné, Karin 8, 57
Ghosh, Amitav 15
global warming 31
GLOF (Glacial Lake Outburst Flood) 12–13, 23; in Sikkim 17–18
Gore, Al 66
Green Buddhism 45
Greening Canadian Mosques 34
Green Muslims 35
Gyatso, Tenzin 46

Haberman, David 6
*hadith* 7, 26
hajj pilgrimage 32
Haluza-DeLay, Randolph 6
Hanh, Thich Nhat 45–46
Harris, Ian 49
Hawai'i 14; traditional beliefs 16–17; watery relations 21
hermeneutics of land 4
Hernandez, Jessica 14–15
Himalayan Buddhist 46
Himalayas: climate change and 12; development 12; rituals to counter the melting glaciers 57–58
Hinduism 13
Holmes-Tagchungdarpa, Amy 14, 46
Ho'omana 16
*How the World's Religions Are Responding to Climate Change* (Veldman) 5

human-environment relations 8
human relatedness with waterways 19–20

Indigenous communities 4, 7, 15–16, 85; and climate change 4, 13–14
Indigenous Himalayan culture 46
Indigenous religions 17, 22; and environmental knowledge 16–17
interdependent origination 46
InterTribal Buffalo Council (ITBC) 22
Islam and climate change 4, 13, 26–27, 83; actions taken 33–34; civil society 34–35; and environmental concern 27–31; political elites 31–34; principle of *'adl* (justice) 28; proclamations and plans 32–33; public opinion 36–40; references in primary sources 28–30; religiously rooted environmental ethics 27–28; secondary religious scholarship 30–31
Islamic Declaration on Global Climate Change 35
Islamic Foundation for Ecology and Environmental Sciences (IFEES) 35
Islamic Relief Worldwide 35

Jenkins, Willis 6
Jordan 40

Kānaka Maoli communities 7, 15, 17, 23
Kanchendzonga 13
Khalid, Fazlun 28
*Khilafah* 27
Kiessling, Christopher 66–67
kilo adaptation 21
Kinchumzongbu chyu 18
Kiowa communities 7, 14–15, 17, 21–22; Sun and 21–22
Kotzé, Louis J. 85
Krech, Shepard 17
Kyoto Protocol, 1997 66

Lal, Vinay 6
Laudato Si 9, 62, 69–72, 83
LaVasseur, Todd 6
Lebanon 40
Lepcha, Charisma 18–19
Lepcha, Kachyo 13
Lhopo (Bhutia) community 7, 13, 16–18, 22
life release 49–52
local worldviews and concerns 2

Loka Initiative 86
Loy, David 48
*lungta* 56

Macy, Joanna 47–49
Maoli, Kānaka 20–21
Maunakea 21
Mayel Liang 18
McMahan, David 47
MENA countries 32–33, 41
Mental Health and Our Changing Climate: Impacts, Implications, and Guidance 86
Ministry of Environment and Climate Change (MMA) 68, 69
*Mizan* 28
mo'o 20
Morocco 34, 40
Mun Bongthing's rituals 18
*Muslim Environmentalism* (Gade) 6
Muslim non-governmental organizations 26
Muslim practitioners 4, 7–8; *see also* Islam and climate change
*mutanchi rongcup rumkup* 18

Nasr, Seyyed Hossein 28, 30
Nepal 56
Nigeria 34
Nobre, Carlos 8, 73
norm entrepreneurs 62–63

OIC *see* Organisation of Islamic Cooperation
OneEarthSangha 44
Organisation of Islamic Cooperation (OIC) 31

*Pachamama* 85
Palestine 40
Pan-Amazonian Church Network (REPAM) 73
Paris Agreement, 2023 32–33, 83
Paris Climate Change Conference 70
personal belief 2
Phrakhru Manas Natheephitak 53
Pope Francis 65, 71
*Propitiation Rite for the Sacred Habitat of the Valley of Rice* 58

Quadresimo Anno 64
Quran 7, 26–28; 17:27 27; 55:7–8 28; 6:141 28; 6:38 29; 45:13 29; 16:14 29

Rabat 33
Rathong River dam project 18–19
Ratnabhava 51
religion 1, 3; and climate change 5; manifestation in community 5, 82–83; understanding 5
religious actors 2
religious communities: and climate change 3–4; conversations 1, 4, 81–82
Resilience in The Anthropocene (RITA) online summit 86
Rinpoche, Patrul 50
Rinzin, Chewang 58
rituals to counter melting glaciers 57–58
Rong (Lepcha) community 7, 12–13, 18, 22

*sadaqah* 28
*sakhkhara* 30
Saudi Arabia 32
Schellnhuber, Hans Joachim 4
scientific communities, conversations 1
SDGs *see* Sustainable Development Goals
Sherpa, Ang Dolma 56–57
Sherpa, Pasang Yangjee 12, 23
Shiu, Henry 52
Sikkim, Himalayas 13, 46, 58; colonialism 14; GLOF (Glacial Lake Outburst Flood) in 17–18; mountains as kin and protectors 17–19
Sikkink, Kathryn 62–63
Silva, Lula da 75
Silva, Marina 67
*Skin jug* ritual 58
Smokey the Bear Sūtra 48
Snyder, Jack 48–49
social justice advocates 8
Stanley, John 44
Stevenson, Daniel 51–52
Stokes, Leah 52
Strain, Charles R. 48
Sustainable Development Goals (SDGs) 70
Synod for the Amazon 72–75

Tan, Gillian 50
Teesta III megadam 12–13
Te Punga Somerville, Alice 16

Theravada Thai society 53
three "Bs" (belief, behavior, and belonging) 37
Tibetan Buddhist cosmological framewqrk 49, 84
Tinker, Tink 16
Treaty of Amazon Cooperation, 1978 66
Treaty of Madrid 66
Tsing, Anna 6
Turkey 32
Turkson, Cardinal 76
Tzu Chi Foundation 55–56

*ummah* 35
*Understanding Climate Change through Religious Lifeworlds* (Haberman) 6
UN Framework Convention on Climate Change 35
United Arab Emirates (UAE) 32
United Nations Climate Change Conference 3: COP21 44, 58, 83; COP26 32; COP27 31
United Nations Conference on the Human Environment 66
United Nations Convention of Climate Change 3
United Nations Permanent Forum on Indigenous Issues 16
Utpala Craft 56

Veldman, Robin Globus 5
*Vinaya* 53

wai 20
Walker, Polly 16
waste *(israf)* 27–29
Weaver, Jace 6
White Jr., Lynn 17
Whyte, Kyle 15–16, 23
*Words of My Perfect Teacher (Kun bzang bla ma'i zhal lung)* 50
World Values Survey 36

Yellowstone River Valley 17
Yemen 40

Zampini, Augusto 74
zero-carbon economies 32
*zhidak* 58
Zimmerman-Liu, Teresa 56

For Product Safety Concerns and Information please contact our EU representative  GPSR@taylorandfrancis.com
Taylor & Francis Verlag GmbH, Kaufingerstraße 24, 80331 München, Germany

www.ingramcontent.com/pod-product-compliance
Lightning Source LLC
Chambersburg PA
CBHW071823230426
43670CB00013B/2550